EXPLORATIONS IN PARALLEL DISTRIBUTED PROCESSING

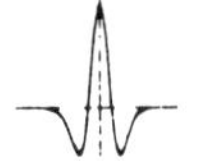

Computational Models of Cognition and Perception

Editors

Jerome A. Feldman
Patrick J. Hayes
David E. Rumelhart

Parallel Distributed Processing: Explorations in the Microstructure of Cognition. Volume 1: Foundations, by David E. Rumelhart, James L. McClelland, and the PDP Research Group

Parallel Distributed Processing: Explorations in the Microstructure of Cognition. Volume 2: Psychological and Biological Models, by James L. McClelland, David E. Rumelhart, and the PDP Research Group

Neurophilosophy: Toward a Unified Science of the Mind-Brain, by Patricia Smith Churchland

Qualitative Reasoning About Physical Systems, edited by Daniel G. Bobrow

Induction: Processes of Inference, Learning, and Memory, by John H. Holland, Keith J. Holyoak, Richard E. Nisbett, and Paul R. Thagard

Production System Models of Learning and Development, edited by David Klahr, Pat Langley, and Robert Neches

Minimal Rationality, by Christopher Cherniak

Vision, Brain, and Cooperative Computation, edited by Michael A. Arbib and Allen R. Hanson

Computational Complexity and Natural Language, by G. Edward Barton, Jr., Robert C. Berwick, and Eric Sven Ristad

Explorations in Parallel Distributed Processing: A Handbook of Models, Programs, and Exercises, by James L. McClelland and David E. Rumelhart

EXPLORATIONS IN PARALLEL DISTRIBUTED PROCESSING

A Handbook of Models, Programs, and Exercises

James L. McClelland
David E. Rumelhart

A Bradford Book

The MIT Press
Cambridge, Massachusetts
London, England

Second printing, 1991

This book was printed and bound in the United States of America

Library of Congress Cataloging-in-Publication Data

McClelland, James L.
Explorations in parallel distributed processing.

Computational models of cognition and perception
"A Bradford book."
Includes bibliographical references.
1. Parallel processing (Electronic computers)
2. Electronic data processing—Distributed processing
I. Rumelhart, David E. II. Title. III. Series.
Q76.5.M295 1989 004'.35 89-13036
ISBN 0-262-63129-6

Contents

Preface

Uses of This Handbook

We have produced this handbook primarily to help others who wish to undertake their own explorations in parallel distributed processing. The handbook can be used in a self-guided program of exploration. We provide tutorials on the theoretical background of all the models considered, as well as simple exercises to help the reader develop a feeling for the basic concepts of PDP and for the simulation models that embody these principles.

The material covered comes largely from the two volumes on parallel distributed processing that grew out of our work with the PDP research group at UCSD (McClelland, Rumelhart, & the PDP Research group, 1986; Rumelhart, McClelland, & the PDP Research Group, 1986). This represents fairly broad coverage, since many prominent workers in the field contributed to these volumes and many of the models built upon the pioneering work of others.

The range of models covered should be sufficiently broad to let this volume serve as a primer in PDP or connectionist modeling. There are some gaps, but we have found that once students have gained some hands-on experience with a subset of models in the field, there is considerable positive transfer, and they are able to assimilate written material about other models better once they have had this experience.

The handbook should prove suitable for courses that serve as a first introduction to PDP or connectionist modeling. We have used this material successfully in courses for juniors, seniors, and graduate students who have had no prior experience in the area. We hope that the coverage of basic theoretical material provided in each chapter will help to free instructors to devote some lectures and discussions to the ideas and motivations behind the models and to introduce material we have not covered here.

We have given answers to the questions posed in the exercises to encourage students to use the questions to test their own understanding of the material and to help anyone who uses the handbook for self-study purposes to stay on track. We have also tried to make it possible for students to explore appplications of the various models to problems they pose for themselves. In some cases, it may even be feasible for students with programming experience to extend the software and explore variants of the models that we have not implemented or even to develop new models of their own. However, our experience has been that most students do not have the time, within a course that lasts a single quarter or semester, to do too much extension of the programs themselves; this is often better left for a follow-up, independent study course in a later term.

An additional goal of the handbook is to make our simulation programs available to researchers who already have some specific research interest in PDP. Over the years we have frequently been asked for copies of our programs by researchers who wished to try our simulation models on their own problems or to modify and extend the models. Judging from our own experience, the software provided with this handbook will prove useful in meeting many of these needs; we find that the **bp** program, in particular, has become a major workhorse that has been adapted by us and many of our graduate students for use in a number of ongoing research projects.

We have retained copyright to the text and accompanying software, but encourage the use and modification of the software for educational and research purposes. We reserve all rights to market the text and software, both source and object code.

Acknowledgments

This book grew out of courses we have taught at Carnegie-Mellon and UCSD over the last three years. At first, we only intended to make the simulation programs we had developed in the course of our research available for students in our home institutions, but the response to the opportunity to learn about PDP by using the programs has been so enthusiastic that it seemed clear there would be a broader audience. It is to our students, then, that we owe the inspiration to produce this handbook.

The specific suggestion to publish a version of the software came from Eric Jenkins, who was at the time (March, 1986) a first-year psychology graduate student at Carnegie-Mellon. In an early conversation with Eric, several basic elements of the philosophy behind the design of the software were laid out.

This handbook also owes a great deal to the students in a course taught by Dave Rumelhart at UCSD in the fall of 1986 and to the students in a course taught by Jay McClelland at Carnegie Mellon in the spring of 1987. These students made do with sometimes buggy versions of the programs and often sketchy drafts of the chapters, and provided us with very useful feedback and suggestions. We would like to mention particularly the useful comments we received from Benoit Mulsant, who caught a number of glaring errors. We also received very useful feedback on a portion of the handbook from Jeremy Roschelle at Berkeley.

Elliot Jaffe, then a research programmer at CMU, played a major role in the development of the user interface that we are now using in our research simulators, as well as in the programs distributed with this handbook. Elliot did an excellent job designing this user interface to our specifications and implementing the first version of the programs, adapting our existing simulators to work with the new interface. Elliot's work was supported by a contract from the Office of Naval Research (N00014-82-C-0374, NR 442a-483). We have done the fine-tuning of the software for educational use, with assistance from Michael Franzini at Carnegie-Mellon. Mike has played a particularly important role in streamlining the PC-specific input/output routines and in finalizing the PC version of the software.

Copyediting and production of the handbook was done at UCSD. We would like to thank Don Norman, director of the Institute for Cognitive Science at UCSD, and Sondra Buffett, administrative director, for their general support and understanding as this project has continued even after both of us have left our former home at UCSD. To Kathy Farrelly at UCSD we owe a very large debt of gratitude. As with the earlier PDP volumes, she has taken care of the copyediting, the preparation of figures, the production of camera-ready copy, and a thousand other things with extraordinary thoughtfulness and efficiency. We would also like to thank Rebecca Duxbury at Carnegie-Mellon for meticulous secretarial assistance in all phases of the preparation of the text.

Finally, we would like to thank Harry and Betty Stanton and their associates at MIT Press for their support. The Press provided the use of a PC for software development as well as financial assistance for the production of camera-ready copy of the text, and has still been able to keep the handbook affordable. We also thank the Stantons for understanding when we did not meet the wildly optimistic target date we gave them for completing this handbook. In retrospect, the eighteen months it has taken seems very good for a project such as this, even though we never dreamed it would take that long when we began.

September 1987

James L. McClelland
PITTSBURGH, PENNSYLVANIA

David E. Rumelhart
PALO ALTO, CALIFORNIA

EXPLORATIONS IN PARALLEL DISTRIBUTED PROCESSING

CHAPTER 1

Introduction

Parallel distributed processing provides a new way of thinking about perception, memory, learning, and thinking—and about basic computational mechanisms for intelligent information processing in general. These new ways of thinking have all been captured in simulation models. Our own understanding of parallel distributed processing (PDP) has come about largely through hands-on experimentation with these models. And, in teaching PDP to others, we have discovered that their understanding is enhanced through the same kind of hands-on simulation experience.

This handbook is intended to help a wider audience gain this kind of experience. It makes many of the simulation models discussed in the two PDP volumes (McClelland, Rumelhart, & the PDP Research Group, 1986; Rumelhart, McClelland, & the PDP Research Group, 1986) available in a form that is both accessible and easy to use. The handbook also provides what we hope are relatively accessible expositions of some of the main mathematical results that underlie the simulation models. And it provides a number of prepared exercises to help the reader begin exploring the simulation programs.

This handbook is intended for use in conjunction with the two PDP volumes, particularly for users new to PDP. However, readers who already have some familiarity with PDP models should find that it is possible to use the simulation programs without referring to the PDP volumes. Most important for readers who are new to PDP is the introduction to the PDP framework found in Chapters 1 to 4 of the first PDP volume. Other chapters in the PDP volumes are less essential, since this handbook generally reviews specific material where relevant. However, the PDP volumes generally delve more deeply into the relevant theoretical and empirical background. Rather than repeat this material, we give pointers throughout the handbook to relevant material from the PDP volumes.

In this chapter, we provide some general information about the use of this handbook. We begin by describing the nature of the software that accompanies this handbook and the hardware you will need to run it. Then we describe what is in each chapter and how the chapters are organized. The final sections of this chapter describe some general conventions and design decisions we have made to help the reader make the best possible use of the handbook and the software that comes with it.

THE PROGRAMS: WHAT IS PROVIDED AND WHAT IS NEEDED TO RUN THEM[1]

This book comes with two 5¼" floppy disks, which contain a set of seven simulation programs as well as auxiliary files that are needed to execute the exercises described in the following chapters. The disks also have utilities for making simple graphs from saved results. In keeping with our goal of maximum accessibility, our programs are compiled for use on IBM PCs or PC-compatible hardware.

We also provide the source code for the programs (written in C) so that users can modify the programs and adapt them for their own purposes. This also makes it possible to copy the programs to more powerful computers, where they may be recompiled.

The minimal hardware requirements to run the programs are

- An IBM PC or PC-compatible computer with two floppy disk drives or one floppy and one Winchester disk drive.
- A standard monochrome 24 line by 80 character display.
- At least 256 kbytes of memory.
- The MS-DOS operating system (Version 2.0 or higher).

If you are using a two-floppy system, you will need several floppy disks. Typically, one floppy will hold two programs in the unpacked, ready-to-run state, together with the relevant auxiliary files. To carry out some of the exercises, you will need to be able to edit files; that is, you will need a text editor. It is also preferrable to have more than 256 kbytes of memory, but this is not essential for running any of the basic exercises.

[1] Please see the software license agreement at the back of this handbook for several important disclaimers an for licensing information concerning the use, modification, and dissemination of this software.

For use on PCs, a math coprocessor (e.g., the 8087) is strongly recommended. All of the programs do extensive floating-point computation and they will run much faster with the coprocessor than without it. On PCs without a coprocessor, some of the exercises in Chapters 2, 5, and 7 will run slower than is optimal for interactive experimentation.

For recompilation on PCs and PC-compatibles, we recommend Microsoft C, which is what we used to produce the PC-executable versions of the programs. We cannot guarantee that other compilers will offer all of the necessary libraries (especially input/output handling, interrupt handling, and math functions such as *exp* and *square root*) or that they will not have bugs we have not encountered. For recompilation on UNIX systems, all that is required is the Portable C compiler, augmented by the CURSES screen-oriented input/output package. On UNIX systems, the programs will run on regular terminals (such as the DEC VT-100 or Zenith Z19) or under terminal-emulation programs on workstations such as IBM-RTs, MicroVAXes, or Suns.

OVERVIEW OF THE HANDBOOK

The handbook consists of seven chapters, including this brief introduction, plus several appendixes. Chapters 2 and 3 are devoted to models that focus primarily on *processing*. In Chapter 2, we describe a class of PDP models called interactive activation and competition models. Models of this kind have been explored by a number of investigators, including ourselves and Grossberg (1976, 1978, 1980). The chapter goes over some of the basic mathematical properties of this sort of model and uses the model to illustrate many of the basic processing capabilities of PDP networks. The chapter then introduces the reader to our simulation programs through the **iac** program. This program implements the interactive activation and competition mechanism and applies it to the problems of memory retrieval, spontaneous generalization, and default assignment using the "Jets and Sharks" example described in Chapter 1 of the PDP volumes (originally from McClelland, 1981). (Henceforth, we will refer to chapters in the PDP volumes by *PDP:N*, where *N* is the chapter number. Chapters 1-13 are in Volume 1; Chapters 14-26 are in Volume 2).

Chapter 2 also serves as an introduction to the package of programs as a whole. Commands that are used in all of the programs are described there, and the exercises are set up to give the reader some general facility with the package, as well as specific experience with interactive activation and competition networks.

Chapter 3 considers several related models that fall into the broad category of constraint satisfaction models. These include what we call the *schema* model (*PDP:14*), the Boltzmann machine (*PDP:7*), and the *harmonium* (*PDP:6*). The chapter uses several of the examples described in the original chapters, allowing the reader to replicate some of the basic simulations that can be found there.

Chapters 4, 5, and 6 describe PDP models of learning. They can be taken up after Chapter 2 if desired. In Chapter 4, two classical learning rules are introduced: the *Hebb* rule and the *delta* rule. These learning schemes are applied in a simulation program called **pa** that implements a classic type of PDP network, the pattern associator, a simple one-layer feedforward network. This class of networks is analyzed in *PDP:9* and *PDP:11*, and is applied to psychological issues, such as the basis of lawful behavior, in *PDP:18* and *PDP:19*.

The network architecture simulated in Chapter 4 involves only a single layer of modifiable connections. Chapter 5 generalizes the architecture to multilayer, feedforward networks, and introduces mechanisms for training *hidden* units—processing units that do not receive any direct input from the outside. This chapter focuses primarily on the **bp** program, which implements the back propagation algorithm introduced in *PDP:8*.

In Chapter 6, two other architectures for learning are considered: the auto-associator and the competitive-learning network. The auto-associator has been used most extensively by James Anderson (1977) and Kohonen (1977); some applications of the auto-associator to issues of learning and memory are discussed in *PDP:17* and *PDP:25*. The competitive-learning scheme (and variants of it) have also been widely studied (e.g., von der Malsberg, 1973, and Grossberg, 1976); our applications of this scheme are described in *PDP:5*. Things are set up so that readers can proceed from the **pa** model described in Chapter 4 directly to any of the other learning models without missing any essential information.

Chapter 7 considers the use of PDP models to simulate psychological phenomena and presents the **ia** simulation program, which implements the interactive activation model of visual word recognition. This model is mentioned in *PDP:1* and *PDP:16*, and is described in detail in two earlier publications (McClelland & Rumelhart, 1981; Rumelhart & McClelland, 1982). This chapter can be taken up immediately after Chapter 2 if desired.

The appendixes provide reference information that will be of use throughout the book. Appendix A describes how to unpack the programs for the PC and how to set up working directories in which to run the exercises. Appendix B provides a command summary, listing all of the commands along with the programs in which they are available, with a brief description of each and pointers to more information. Appendix C describes the construction of files containing network and display specifications for use with the various programs. Appendix D explains how to use the utility programs that are supplied for making simple graphs. Appendix E provides some feedback on the outcome and interpretation of selected exercises. Appendix F provides an overview of the source code for readers with a background in C who wish to modify and recompile the programs. Appendix G explains how to recompile the programs for various computers.

ORGANIZATION OF EACH CHAPTER

Each chapter begins with a brief overview, followed by one or more parts, each devoted to a model or a set of related models. Each part begins with a theoretical background section that is designed to provide an accessible introductory presentation of the relevant mathematical background for the models described in this part of the chapter. After the background section, one or more related models are presented. Each model begins with a description of the assumptions of the model and any variations on these assumptions that will be considered. This is followed by a description of the implementation of the essential routines that carry out the key computations prescribed by the model and a description of how the computer simulation of the model is to be run. The description of how the model is run consists of an overview explaining basically how the simulation is to be used, followed by a list of the commands and variables that need be understood to use the simulation program that implements the model. Finally, the presentation of each model ends with a series of exercises, the first of which is used as an example to illustrate in detail how to run the model. The exercises are generally accompanied by hints, which are intended to make sure the reader knows what is needed to carry out the exercise, both in terms of the commands needed to get the program to do what is necessary and in terms of any more conceptual points that might be relevant.

MODELS AND PROGRAMS

In general, the relation of models to programs may be many to one. That is, more than one model may be implemented by the same program; the different models are implemented by means of switches that alter the program's behavior. This makes more efficient use of disk space and cuts down some on the number of different programs that need to be learned about. Furthermore, the programs generally make use of the same interface and display routines, and most of the commands are the same from one program to the next.

In view of the similarity between the simulation models, the information that is given when each new program is introduced is restricted primarily to what is new. Readers who wish to dive into the middle of the book, then, may find that they need to refer back to commands or features that were introduced earlier. The command summary in Appendix B should help make this as painless as possible.

SOME GENERAL CONVENTIONS AND CONSIDERATIONS

In planning this handbook, we had to make some design decisions and to adopt some fairly arbitrary conventions. Here we will describe some of the general conventions that are used in the book and in the computer programs.

Mathematical Notation

We have adopted a mathematical notation that is internally consistent within this handbook and that facilitates translation between the description of the models in the text and the conventions used to access variables in the programs. Unfortunately, this means that the notation is not always consistent with that introduced in the relevant chapters of the PDP volumes. Here follows an enumeration of the key features of the notation system we have adopted. We begin with the conventions we have used in writing equations to describe models and in explicating their mathematical background.

Scalars. Scalar (single-valued) variables are given in italic typeface. The names of parameters are chosen to be mnemonic words or abbreviations where possible. For example, the decay parameter is called *decay*.

Vectors. Vector (multivalued) variables (e.g., the vector of activations of a set of units) are given in boldface; for example, the external input pattern is called **extinput**. An element of such a vector is given in italic typeface with a subscript. Thus, the ith element of the external input is denoted $extinput_i$. Vectors are often members of larger sets of vectors; in this case, a whole vector may be given a subscript. For example, the ith input pattern in a set of patterns would be denoted $\mathbf{ipattern}_i$.

Weight matrices. Matrix variables are given in uppercase boldface; for example, a weight matrix might be denoted **W**. An element of a weight matrix is given in lowercase italic, subscripted first by the row index and then by the column index. The row index corresponds to the index of the receiving unit, and the column index corresponds to the index of the sending unit. Thus the weight to unit i from unit j would be found in the jth column of the ith row of the matrix, and is written w_{ij}.

Counting. We follow the C language convention and count from 0. Thus if there are n elements in a vector, the indexes run from 0 to $n-1$. Time is a bit special in this regard. Time 0 (t_0) is the time before processing begins; the state of a network at t_0 can be called its "initial state." Time counters are incremented as soon as processing begins within each time step.

Pseudo-C Code

In the chapters, we occasionally give pieces of computer code to illustrate the implementation of some of the key routines in our simulation programs. The examples are written in "pseudo-C"; details such as declarations are left out. Note that the pseudocode printed in the text for illustrating the implementation of the programs is generally not identical to the actual source code; the program examples are intended to make the basic characteristics of the implementation clear rather than to clutter the reader's mind with the details and speed-up hacks that are to be found in the actual programs.

Several features of C need to be understood to read the pseudo-C code. These are listed below.

Comments. Comments in C programs begin with "/*" and end with "*/". Thus the following would be treated as a comment by the compiler:

```
/* This is a comment */
```

We use this convention to introduce comments into the pseudocode so that the code is easier for you to follow.

If statements. Often in the pseudocode we will make use of *if* statements. These should be fairly self-explanatory in most cases. Sometimes, however, we use the form:

```
if (x)
```

where x is some variable. This expression means "if the value of x is not equal to 0." For flag variables, which have the value 0 when off, this corresponds to "if the flag is on." Another notation for the same thing would be

```
if (x != 0)
```

The exclamation point means "not."

Semicolons and curly braces. Semicolons and curly braces are key features of C syntax. The semicolon is used to terminate a statement. Open- ("{") and close- ("}") curly braces are used to group statements together to be treated as a single statement. Thus

```
if (expression) statement;
```

can be expanded to

```
if (expression) {
        statement;
        statement;
        . . .
}
```

Curly braces are also used to bracket the body of a subroutine.

C loop constructs. In our programs the typical loop construct is the *for* loop. *For* loops look like this:

```
for (i = 0; i < n ; i++) {
        statements;
}
```

The parentheses contain three special statements called the initialization statement, the end test, and the incrementation statement, respectively. In this example we initialize the variable i to 0. The end test is a statement that is evaluated at the beginning of each pass, before executing the statements in the loop; if the end test fails, control passes to the statement after the closing curly brace. In this case, the end test fails when the value of i is greater than or equal to n. The incrementation statement is executed at the end of each pass through the loop before the end test is done to see if the loop will be executed again. The notation

```
i++
```

indicates incrementation of the index i by 1. Thus, the above expression executes the statements enclosed in the curly braces once for each value of i from 0 through $n-1$, for a total of n passes through the loop.

Array indexes. Array indexes in C are enclosed in square brackets, with the fastest moving array element in the right-most position. Thus,

```
w[i][j]
```

refers to the element in the ith row and jth column of the array. Contiguous elements in the array are the contiguous members of the same row. The above notation, then, is the notation that refers to the weight to unit i from unit j, and corresponds to w_{ij}.

Incrementing and related constructs. We have already described the notation for incrementing a variable by 1. To increment a variable by the value of an arbitrary expression, the notation is

<*variable*> += <*expression*>;

Thus,

x += *a***b*;

means "increment *x* by *a* times *b*." There are also related expressions for decrementing (-=), multiplying (*=), and dividing (/=). Thus,

x /= 7;

means "set *x* to *x* divided by 7."

The reader should note that there are a number of features and conventions in C that are exploited extensively in the actual code that can only be understood with some background in this language. Users who do not know C will not be able to interpret the actual code. The place to go for this background is the book *The C Programming Language* by Kernighan and Ritchie (1978).

Computer Programs and User Interface

Our goals in writing the programs were to make them both as flexible as possible and as easy as possible to use, especially for running the core exercises discussed in each chapter of this book. We have achieved these somewhat contradictory goals as follows. Flexibility is achieved by allowing the user to specify the details of the network configuration and of the layout of the displays shown on the screen at run time, via files that are read and interpreted by the program. Ease of use is achieved by providing the user with the files to run the core exercises and by keeping the command interface and the names of variables consistent from program to program wherever possible. Full exploitation of the flexibility provided by the programs requires the user to learn how to construct network configuration files and display configuration (or template) files, but this is only necessary when the user wishes to apply a program to some new problem of his or her own.

Another aspect of the flexibility of the programs is their permissiveness. In general, we have allowed the user to examine and set as many of the variables in each program as possible, including basic network configuration variables that should not be changed in the middle of a run. The worst that can happen is that the programs will crash under these circumstances; it is, therefore, wise not to experiment with changing them if losing the state of a program would be costly.

BEFORE YOU START

Before you dive into your first PDP model, we would like to offer both an exhortation and a disclaimer. The exhortation is to take what we offer here, not as a set of fixed tasks to be undertaken, but as raw material for your own explorations. We have presented the material following a structured plan, but this does not mean that you should follow it any more than you need to to meet your own goals. We have learned the most by experimenting with and adapting ideas that have come to us from other people rather than from sticking closely to what they have offered, and we hope that you will be able to do the same thing. The flexibility that has been built into these programs is intended to make exploration as easy as possible, and we provide source code so that users can change the programs and adapt them to their own needs and problems as they see fit.

The disclaimer is that we cannot be sure the programs are perfectly bug free. They have all been extensively tested and they work for the core exercises; but it is possible that some users will discover problems or bugs in undertaking some of the more open-ended extended exercises. If you have such a problem, we hope that you will be able to find ways of working around it as much as possible or that you will be able to fix it yourself. In any case, please let us know of the problems you encounter.[2] While we cannot offer to provide consultation or fixes for every reader who encounters a problem, we will use your input to improve the package for future users.

---

[2] Send bug reports, problems, and suggestions to PDP Software Inquiries, c/o Texts Manager, MIT Press, 55 Hayward Street, Cambridge, MA 02142. Enclose a stamped, self-addressed envelope, and we will send an acknowledgment of your problem along with accumulated advice, fixes, work-arounds, and information about availability of subsequent releases.

CHAPTER 2

Interactive Activation and Competition

Our own explorations of parallel distributed processing began with the use of interactive activation and competition mechanisms of the kind we will examine in this chapter. We have used these kinds of mechanisms to model visual word recognition (McClelland & Rumelhart, 1981; Rumelhart & McClelland, 1982) and to model the retrieval of general and specific information from stored knowledge of individual exemplars (McClelland, 1981), as described in *PDP:1*. In this chapter, we describe some of the basic mathematical observations behind these mechanisms, and then we introduce the reader to a specific model that implements the retrieval of general and specific information using the "Jets and Sharks" example discussed in *PDP:1* (pp. 25-31). (The interactive activation model of word perception is presented in Chapter 7.)

After describing the specific model, we will introduce the program in which this model is implemented: the **iac** program (for interactive activation and competition). The description of how to use this program will be quite extensive; it is intended to serve as a general introduction to the entire package of programs since the user interface and most of the commands and auxiliary files are common to all of the programs. After describing how to use the program, we will present several exercises, including an opportunity to work with the Jets and Sharks example and an opportunity to explore an interesting variant of the basic model, based on dynamical assumptions used by Grossberg (e.g., Grossberg, 1978).

BACKGROUND

The study of interactive activation and competition mechanisms has a long history. They have been extensively studied by Grossberg. A useful introduction to the mathematics of such systems is provided in Grossberg

(1978). Related mechanisms have been studied by a number of other investigators, including Levin (see Levin, 1976), whose work was instrumental in launching our exploration of PDP mechanisms.

An interactive activation and competition network (hereafter, *IAC network*) consists of a collection of processing units organized into some number of competitive pools. There are excitatory connections among units in different pools and inhibitory connections among units within the same pool. The excitatory connections between pools are generally bidirectional, thereby making the processing *interactive* in the sense that processing in each pool both influences and is influenced by processing in other pools. Within a pool, the inhibitory connections are usually assumed to run from each unit in the pool to every other unit in the pool. This implements a kind of competition among the units such that the unit or units in the pool that receive the strongest activation tend to drive down the activation of the other units.

The units in an IAC network take on continuous activation values between a maximum and minimum value, though their output—the signal that they transmit to other units—is not necessarily identical to their activation. In our work, we have tended to set the output of each unit to the activation of the unit minus the *threshold* as long as the difference is positive; when the activation falls below threshold, the output is set to 0. Without loss of generality, we can set the threshold to 0; we will follow this practice throughout the rest of this chapter. A number of other output functions are possible; Grossberg (1978) describes a number of other possibilities and considers their various merits.

The activations of the units in an IAC network evolve gradually over time. In the mathematical idealization of this class of models, we think of the activation process as completely continuous, though in the simulation modeling we approximate this ideal by breaking time up into a sequence of discrete steps.

Units in an IAC network change their activation based on a function that takes into account both the current activation of the unit and the net input to the unit from other units or from outside the network. The net input to a particular unit (say, unit i) is the same in almost all the models described in this volume: it is simply the sum of the influences of all of the other units in the network plus any external input from outside the network. The influence of some other unit (say, unit j) is just the product of that unit's output, $output_j$, times the strength or weight of the connection to unit i from unit j. Thus the net input to unit i is given by

$$net_i = \sum_j w_{ij}\, output_j + extinput_i. \tag{1}$$

In the IAC model, $output_j = [a_j]^+$. Here, a_j refers to the activation of unit j, and the expression $[a_j]^+$ has value a_j for all $a_j > 0$; otherwise its value is 0. The index j ranges over all of the units with connections to unit i. In general the weights can be positive or negative, for excitatory or inhibitory connections, respectively.

Once the net input to a unit has been computed, the resulting *change* in the activation of the unit is as follows:

$$\text{If } (net_i > 0),\quad \Delta a_i = (max - a_i) net_i - decay(a_i - rest).$$

$$\text{Otherwise},\quad \Delta a_i = (a_i - min) net_i - decay(a_i - rest).$$

Note that in this equation, *max*, *min*, *rest*, and *decay* are all parameters. In general, we choose $max = 1$, $min \leqslant rest \leqslant 0$, and *decay* between 0 and 1. Note also that a_i is assumed to start, and to stay, within the interval [*min, max*].

Suppose we imagine the input to a unit remains fixed and examine what will happen across time in the equation for Δa_i. For specificity, let's just suppose the net input has some fixed, positive value. Then we can see that Δa_i will get smaller and smaller as the activation of the unit gets greater and greater. For some values of the unit's activation, Δa_i will actually be negative. In particular, suppose that the unit's activation is equal to the resting level. Then Δa_i is simply $(max - rest) net_i$. Now suppose that the unit's activation is equal to *max*, its maximum activation level. Then Δa_i is simply $(-decay)(max - rest)$. Between these extremes there is an equilibrium value of a_i, at which Δa_i is 0. We can find what the equilibrium value is by setting Δa_i to 0 and solving for a_i:

$$0 = (max - a_i) net_i - decay(a_i - rest)$$

$$- (max)(net_i) + (rest)(decay) - a_i(net_i + decay)$$

$$a_i = \frac{(max)(net_i) + (rest)(decay)}{net_i + decay}. \tag{2}$$

Using $max = 1$ and $rest = 0$, this simplifies to

$$a_i = \frac{net_i}{net_i + decay}. \tag{3}$$

What the equation indicates, then, is that the activation of the unit will reach equilibrium when its value becomes equal to the ratio of the net input divided by the net input plus the decay. Note that in a system where the activations of other units—and thus of the net input to any particular unit—are also continually changing, there is no guarantee that activations will ever completely stabilize—although in practice, as we shall see, they often seem to.

Equation 3 indicates that the equilibrium activation of a unit will always increase as the net input increases; however, it can never exceed 1 (or, in the general case, *max*) as the net input grows very large. Thus, *max* is indeed the upper bound on the activation of the unit. For small values of the net input, the equation is approximately linear since $x/(x+c)$ is approximately equal to x/c for x small enough.

We can see the decay term in Equation 3 as acting as a kind of restoring force that tends to bring the activation of the unit back to 0 (or to *rest*, in the general case). The larger the value of the decay term, the stronger this force is, and therefore the lower the activation level will be at which the activation of the unit will reach equilibrium. Indeed, we can see the decay term as scaling the net input if we rewrite the equation as

$$a_i = \frac{net_i/decay}{(net_i/decay)+1}. \tag{4}$$

When the net input is equal to the decay, the activation of the unit is 0.5 (in the general case, the value is $(max+rest)/2$). Because of this, we generally scale the net inputs to the units by a strength constant that is equal to the decay. Increasing the value of this strength parameter or decreasing the value of the decay increases the equilibrium activation of the unit.

In the case where the net input is negative, we get entirely analogous results:

$$a_i = \frac{(min)(net_i) - (decay)(rest)}{net_i - decay}. \tag{5}$$

Using $rest = 0$, this simplifies to

$$a_i = \frac{(min)(net_i)}{net_i - decay} \tag{6}$$

This equation is a bit confusing because net_i and *min* are both negative quantities. It becomes somewhat clearer if we use *amin* (the absolute value of *min*) and $anet_i$ (the absolute value of net_i). Then we have

$$a_i = -\frac{(amin)(anet_i)}{anet_i + decay}. \tag{7}$$

What this last equation brings out is that the equilibrium activation value obtained for a negative net input is scaled by the magnitude of the minimum (*amin*). Inhibition both acts more quickly and drives activation to a lower final level when *min* is farther below 0.

How Competition Works

So far we have been considering situations in which the net input to a unit is fixed and activation evolves to a fixed or stable point. The interactive activation and competition process, however, is more complicated than this because the net input to a unit changes as the unit and other units in the same pool simultaneously respond to their net inputs. One effect of this is to amplify differences in the net inputs of units. Consider two units a and b that are in competition, and imagine that both are receiving some excitatory input from outside but that the excitatory input to a (e_a) is stronger than the excitatory input to b (e_b). Let γ represent the strength of the inhibition each unit exerts on the other. Then the net input to a is

$$net_a = e_a - \gamma(output_b) \tag{8}$$

and the net input to b is

$$net_b = e_b - \gamma(output_a). \tag{9}$$

As long as the activations stay positive, $output_i = a_i$, so we get

$$net_a = e_a - \gamma a_b \tag{10}$$

and

$$net_b = e_b - \gamma a_a. \tag{11}$$

From these equations we can easily see that b will tend to be at a disadvantage since the stronger excitation to a will tend to give a a larger initial activation, thereby allowing it to inhibit b more than b inhibits a. The end result is a phenomenon that Grossberg (1976) has called "the rich get richer" effect: Units with slight initial advantages, in terms of their external inputs, amplify this advantage over their competitors.

Resonance

Another effect of the interactive activation process has been called "resonance" by Grossberg (1978). If unit a and unit b have mutually excitatory connections, then once one of the units becomes active, they will tend to keep each other active. Activations of units that enter into such mutually excitatory interactions are therefore sustained by the network, or "resonate" within it, just as certain frequencies resonate in a sound chamber. In a network model, depending on parameters, the resonance can sometimes be strong enough to overcome the effects of decay. For example, suppose that two units, a and b, have bidirectional, excitatory connections with strengths of $2 \times decay$. Suppose that we set each unit's activation at 0.5 and then remove all external input and see what happens. The activations will stay at 0.5 indefinitely because

$$\begin{aligned}\Delta a_a &= (1 - a_a)net_a - (decay)a_a \\ &= (1 - 0.5)(2)(decay)(0.5) - (decay)(0.5) \\ &= (0.5)(2)(decay)(0.5) - (decay)(0.5) \\ &= 0.\end{aligned}$$

Thus, IAC networks can use the mutually excitatory connections between units in different pools to sustain certain input patterns that would otherwise decay away rapidly in the absence of continuing input. The interactive activation process can also activate units that were not activated directly by external input. We will explore these effects more fully in the exercises that are given later.

Hysteresis and Blocking

Before we finish this consideration of the mathematical background of interactive activation and competition systems, it is worth pointing out that the rate of evolution towards the eventual equilibrium reached by an IAC network, and even the state that is reached, is affected by initial conditions. Thus if at time 0 we force a particular unit to be on, this can have the effect of slowing the activation of other units. In extreme cases, forcing a unit to be on can totally block others from becoming activated at all. For example, suppose we have two units, a and b, that are mutually inhibitory, with inhibition parameter γ equal to 2 times the strength of the decay, and suppose we set the activation of one of these units—unit a—to 0.5. Then the net input to the other—unit b—at this point will be $(-0.5)(2)(decay) = -decay$. If we then supply external excitatory input to the two units with strength equal to the decay, this will maintain the activation of unit a at 0.5 and will fail to excite b since its net input will be 0. The external input to b is thereby blocked from having its normal effect. If external input is withdrawn from a, its activation will gradually decay (in the absence of any strong resonances involving a) so that b will gradually become activated. The first effect, in which the activation of b is completely blocked, is an extreme form of a kind of network behavior known as *hysteresis* (which means "delay"); prior states of networks tend to put them into states that can delay or even block the effects of new inputs.

Because of hysteresis effects in networks, various investigators have suggested that new inputs may need to begin by generating a "clear signal," often implemented as a wave of inhibition. Such ideas have been proposed by various investigators as an explanation of visual masking effects (see, e.g., Weisstein, Ozog, & Szoc, 1975) and play a prominent role in Grossberg's theory of learning in neural networks (see Grossberg, 1980).

Grossberg's Analysis of Interactive Activation and Competition Processes

Throughout this section we have been referring to Grossberg's studies of what we are calling interactive activation and competition mechanisms. In fact, he uses a slightly different activation equation than the one we have presented here (taken from our earlier work with the interactive activation model of word recognition). In Grossberg's formulation, the excitatory and inhibitory inputs to a unit are treated separately. The excitatory input (e) drives the activation of the unit up toward the maximum, whereas the inhibitory input (i) drives the activation back down toward the minimum. As in our formulation, the decay tends to restore the activation of the unit to its resting level.

$$\Delta a = (max - a)e - (a - min)i - decay(a - rest). \tag{12}$$

Grossberg's formulation has the advantage of allowing a single equation to govern the evolution of processing instead of requiring an *if* statement to intervene to determine which of two equations holds. It also has the characteristic that the direction the input tends to drive the activation of the unit is affected by the current activation. In our formulation, net positive input tends always to excite the unit and net negative input tends always to inhibit it. In Grossberg's formulation, the input is not lumped together in this way. As a result, the effect of a given input (particular values of e and i) can be excitatory when the unit's activation is low and inhibitory when the unit's activation is high. Furthermore, at least when *min* has a relatively small absolute value compared to *max*, a given amount of inhibition will tend to exert a weaker effect on a unit starting at rest. To see this, we will simplify and set $max = 1.0$ and $rest = 0.0$. By assumption, the unit is at rest so the above equation reduces to

$$\Delta a = (1)(e) - (amin)(i) \tag{13}$$

where *amin* is the absolute value of *min* as above. This is in balance only if $i = e/amin$.

Our use of the net input rule was based primarily on the fact that we found it easier to follow the course of simulation events when the balance of excitatory and inhibitory influences was independent of the activation of the receiving unit. However, this by no means indicates that our formulation is superior computationally. Therefore we have made Grossberg's update rule available as an option in the **iac** program.

THE IAC MODEL

The IAC model provides a discrete approximation to the continuous interactive activation and competition processes that we have been considering up to now. We will consider two variants of the model: one that follows the interactive activation dynamics from our earlier work and one that follows the formulation offered by Grossberg.

Architecture

The IAC model consists of several units, divided into *pools*. In each pool, all the units are mutually inhibitory. Between pools, units may have excitatory connections. The model assumes that these connections are bidirectional, so that whenever there is an excitatory connection from unit i to unit j, there is also an excitatory connection from unit j back to unit i.

Visible and Hidden Units

In an IAC network, there are generally two classes of units: those that can receive direct input from outside the network and those that cannot. The first kind of units are called *visible* units; the latter are called *hidden* units. Thus in the IAC model the user may specify a pattern of inputs to the visible units, but by assumption the user is not allowed to specify external input to the hidden units; their net input is based only on the outputs from other units to which they are connected.

Activation Dynamics

Time is not continuous in the IAC model (or any of our other simulation models), but is divided into a sequence of discrete steps, or *cycles*. Each cycle begins with all units having an activation value that was determined at the end of the preceding cycle. First, the inputs to each unit are computed. Then the activations of the units are updated. The two-phase procedure ensures that the updating of the activations of the units is effectively synchronous; that is, nothing is done with the new activation of any of the units until all have been updated.

The discrete time approximation can introduce instabilities if activation steps on each cycle are large. This problem is eliminated, and the approximation to the continuous case is generally closer, when activation steps are kept small on each cycle.

Parameters

In the IAC model there are several parameters under the user's control. Most of these have already been introduced. They are

max
: The maximum activation parameter.

min
: The minimum activation parameter.

rest
: The resting activation level to which activations tend to settle in the absence of external input.

decay
: The decay rate parameter, which determines the strength of the tendency to return to resting level.

estr
: This parameter stands for the strength of external input (i.e., input to units from outside the network). It scales the influence of external signals relative to internally generated inputs to units.

alpha
: This parameter scales the strength of the excitatory input to units from other units in the network.

gamma
: This parameter scales the strength of the inhibitory input to units from other units in the network.

In general, it would be possible to specify separate values for each of these parameters for each unit. The IAC model does not allow this, as we have found it tends to introduce far too many degrees of freedom into the modeling process. However, the model does allow the user to specify strengths for the individual connection strengths in the network.

IMPLEMENTATION

The IAC model is implemented by the **iac** program. This program, like all of our simulation programs, is written in C. The program consists of several parts: the command interpreter, the display package, the network configuration package, the patterns package, and the core routines of the model. In describing the implementation of this and other models, we will focus our attention on the core routines, but here we will briefly describe the rest of the package so that the reader has some pointers to understanding what is going on. More detailed implementation information is provided in Appendix F, which serves as a guide for readers who wish to actually explore and possibly alter the source code itself. Here follow brief descriptions of the various noncore parts of the program.

The Command Interpreter

The command interpreter is a set of subroutines that reads commands, either from a start-up file when the program is first called or from the keyboard while the program is running. There is also a facility that allows the user to direct the command interpreter to read and execute a sequence of commands found in a file. The command interpreter works by looking up commands it encounters in a large table of commands and executing the subroutine that is found in the table associated with the command. We will explain how to use the command interpreter in the section "Running the Program" later in this chapter.

The Display Package

The display package is a set of routines that manages the 24×80 character display screen. One set of routines is used to read a file called the *template* file when the program is first called. The information in this file is used to set up a set of *templates*, or display objects, and to indicate what each template contains and where on the screen it should be displayed. Another set of routines is used to do the actual displaying; these are commands that can be issued either by the user directly or from other parts of the program.

The Network Configuration Package

This package consists of a set of routines that is used in configuring the program for a particular application. The routines read commands from a file called the *network configuration* file, or the *network* file for short, and use these commands to set up arrays for the units and the weights, to specify initial values for the connections, and to indicate whether connections are modifiable or not.

The Patterns Package

This package consists of a set of routines that is used to read in a set of patterns for use as inputs to the model. These routines read a file called the *pattern* file. Some of the programs—among them, the **iac** program—can be run without reading in a file full of patterns, but the package is available for use if desired.

The Core Routines

Beyond the routines just mentioned is a set of *core* routines that implements the activation and competition processes described earlier. The routines are simple and make up a rather small part of the program. Here we explain the basic structure of the core routines used in the **iac** program.

getinput. This routine is used to specify which of the units in the network will receive external input. The routine prompts the user for names or numbers of units and for corresponding external input values, after first allowing the external inputs to be cleared to all zeros if desired. Note that this does not actually start the process of sending inputs to the units; it simply says which units should receive inputs and how strong they should be when the process actually starts.

reset. This routine is used to reset the activations of units to their resting levels and to reset the time—the current cycle number—back to 0. All other relevant variables are cleared, and the display is updated to show the initial state of the network before processing begins.

cycle. This routine is the basic routine that is used in running the model. It carries out a number of processing cycles, as determined by the program control variable *ncycles*. On each cycle, two routines are called: *getnet* and *update*. At the end of each cycle, the program checks to see whether the display is to be updated and whether to pause so the user can examine the new state (and possibly terminate processing). At the end of *ncycles* of processing, the display is updated if it has not been updated on every cycle. The routine looks like this:

```
cycle() {

  for (cy = 0; cy < ncycles; cy++) {
    cycleno++;
    getnet();
    update();

/* what follows is concerned with
   pausing and updating the display */
    if (step_size == CYCLE) {
      update_display();
```

```
        if (single_step) {
          if (contin_test() == BREAK) break;
        }
      }
    }
    if (step_size > CYCLE) {
      update_display();
    }
}
```

The *getnet* and *update* routines are somewhat different for the standard version and Grossberg version of the program. We first describe the standard versions of each, then turn to the Grossberg versions.

Standard getnet. The standard *getnet* routine computes the net input to each unit. The net input consists of three things: the external input, scaled by *estr*; the excitatory input from other units, scaled by *alpha*; and the inhibitory input from other units, scaled by *gamma*. For each unit, the *getnet* routine first accumulates the excitatory and inhibitory inputs from other units, then scales the inputs and adds them to the scaled external input to obtain the net input.

Whether a connection is excitatory or inhibitory is determined by its sign. Thus if w_{ij} is positive, $w_{ij}a_j$ is added into the excitation term of unit i. If w_{ij} is negative, $w_{ij}a_j$ is added into the inhibition term of unit i. These operations are only performed if the activation of the sending unit is greater than 0. The code that implements these calculations is as follows:

```
getnet() {

  for (i = 0; i < nunits; i++) {
    excitation[i] = inhibition[i] = 0;

    for (j = 0; j < nunits; j++) {
      if (activation[j] > 0) {
        if (w[i][j] > 0) {
          excitation[i] += weight[i][j]*activation[j];
        }
        else if (w[i][j] < 0) {
          inhibition[i] += weight[i][j]*activation[j];
        }
      }
    }
    netinput[i] = estr*extinput[i] + alpha*excitation[i]
                        + gamma*inhibition[i];
  }
}
```

Standard update. The *update* routine increments the activation of each unit, based on the net input and the existing activation value. Here is what it looks like:

```
update() {

  for (i = 0; i < nunits; i++) {
    if (netinput[i] > 0) {
      activation[i] += (max - activation[i])*netinput[i]
                         - decay*(activation[i] - rest);
    }
    else {
      activation[i] += (activation[i]-min)*netinput[i]
                         - decay*(activation[i] - rest);
    }
    if (activation[i] > max) activation[i] = max;
    if (activation[i] < min) activation[i] = min;
  }
}
```

The last two conditional statements are included to guard against the anomalous behavior that would result if the user had set the *estr*, *istr*, and *decay* parameters to values that allow activations to change so rapidly that the approximation to continuity is seriously violated and activations have a chance to escape the bounds set by the values of *max* and *min*.

Grossberg versions. The Grossberg versions of these two routines are structured like the standard versions. In the *getnet* routine, the only difference is that the net input to each unit is not computed; instead, the excitation and inhibition are scaled by *alpha* and *gamma*, respectively, and scaled external input is added to the excitation if it is positive or is added to the inhibition if it is negative:

```
excitation[i]*= alpha*excitation[i];
inhibition[i]*= gamma*inhibition[i];
if (extinput[i] > 0) excitation[i] += estr*extinput[i];
else if (extinput[i] < 0)
  inhibition[i] += estr*extinput[i];
```

In the *update* routine the two different versions of the standard activation rule are replaced by a single expression. The routine then becomes

```
update() {

  for (i = 0; i < nunits; i++) {
    activation[i] += (max - activation[i])*excitation[i]
                      + (activation[i] - min)*inhibition[i]
                        - decay*(activation[i] - rest);
```

```
        if (activation[i] > max) activation[i] = max;
        if (activation[i] < min) activation[i] = min;
    }
}
```

The reader may have noticed that the main computational loops of the program make no explicit mention of the IAC network architecture, in which the units are organized into competitive (inhibitory) pools and in which excitatory connections are assumed to be bidirectional. These architectural constraints are imposed in the *network* file. In fact, the **iac** program can implement any of a large variety of network architectures, including many that violate the architectural assumptions of the IAC framework.

As these examples illustrate, the core routines of this model—indeed, of all of our models—are extremely simple. Actually, some complexity has been suppressed, but not much. What makes the programs rather complex is all of the auxiliary routines.

RUNNING THE PROGRAM

Starting Up

To run the **iac** program, it is first necessary to set up a working directory containing the relevant files. An explanation of how this is done is given in Appendix A. Here we assume that you have created a working directory for **iac** and that you have positioned yourself in that directory. To execute the program, you would enter the following:

iac <templatefile> <startupfile>

Note that any commands entered either inside or outside of our program must be terminated by pressing the *return* or *enter* key. We adopt the convention of giving variables that must be replaced by specific values inside of angle brackets. Thus *<templatefile>* must be replaced by the name of a specific template file, and *<startupfile>* must be replaced by the name of a specific start-up file. By convention, the names of template files end with the extension *.tem* and the names of start-up files end with the extension *.str*. Henceforth we will refer to the template file as the *.tem* file, and the start-up file as the *.str* file.

The program will run without a template file or a start-up file being given, but the template file is necessary to tell the program what to display and where to display it; without one, there will be no display on the screen. The first argument to the program is always interpreted as a template file name, so the program will misinterpret the *.str* file if the *.tem* file is left out. To prevent this, the program may be run with a single "–" in place of the template file name:

iac – <startupfile>

The *.str* file can be omitted without any ill effects. In general this file contains commands that initialize the network configuration and set the values of various parameters of the model. These can all be entered directly by the user once the program has started to run. The two things that are special about the *.str* file is that the commands in it are executed without printing anything to the screen and that errors encountered in the *.str* file cause the program to terminate immediately, with an error message printed to the screen. The *.str* file can contain any commands the user wishes to put in it, including commands to run the program, save output, and quit. This allows programs to be run in background mode on UNIX systems, using a script of commands from the *.str* file.

Assuming the *.str* file is processed without error and without encountering a *quit* command, the program will present a display containing an **iac**: prompt on line 0, a menu listing commands that may be entered on lines 1 through 4, and a display of the current state of the network. From this point on, the user may enter commands to the program via the command interface.

Entering Commands via the Command Interface

It is useful to think of commands as being entered one per line, with spaces separating the command from its various arguments. For example, the **iac** program provides a command that allows the user to display any of the various display chunks or *templates* that have been specified in the *.tem* file. This command is given by entering

iac: *disp <template>*

where *<template>* is the name of any template specified in the *.tem* file. Note that in this and subsequent examples, we display the prompt typed by the computer in **bold**, with the response from the user in *italic*. Also note that the user interface is case sensitive, and command names are in lowercase throughout. In the exercises we capitalize the first letter of some of the unit names; otherwise everything is lowercase.

A nice feature of the command interface is that it will generally prompt you with possible options should you wish to see them. At the top level, the program always provides a list of the commands that can be entered.

To see lists of options that are specifiable within a given command, enter the command name by itself. Thus if you enter

iac: *disp*

the program comes back with the list of possible continuations of the display command and a revised prompt,

iac: disp /

indicating that you may now enter the continuation you want. If you just want to look at the list of possible continuations, you can type *return* (press the return or enter key), and the program will return to the top level. Alternatively, you may enter one of the available continuations. If further input is required, you will be prompted for it; you may type *return* at almost any time, and the program will revert to the top-level prompt, awaiting a new command input.

One may think of the commands available in the program as consisting of a command name, followed by one or more *specifiers*, followed finally by one or more *arguments*. The specifiers indicate which particular one of several specific commands you wish to execute, and the arguments are the parameters of the command itself. Commands or specifiers that must be followed by further specifiers are terminated in the menus by a "/" character. The other commands or specifiers do not require further specifiers, though the user will generally be prompted for additional arguments. The *display* command is an example of the former type of command: It requires a further specifier indicating which template to display. The *log* command is an example of the latter type. It is used to control the storing of a log of the activity of the network in a file. This command requires a file name argument.

In all cases, the entire command, including the command name, the specifiers, and the arguments, may be entered as a single line. Alternatively, the user may enter any part of a command and be prompted for the possible continuations of it.

A further feature of the command interface is that commands and command specifiers do not have to be entered in their entirety; instead, it is only necessary to type enough of the beginning of the command or specifier to uniquely distinguish it from all other available alternatives. Thus,

iac: *lo foo.log*

is sufficient to open a log file called *foo.log*, given that there are no other commands accessible at the top level that begin with the characters *lo*.

The command interpreter prints an error message if the command string entered is not consistent with any of the available commands. The message is available for a short period (a few seconds), then the command interpreter returns to the top level, ignoring the remainder of a command line on which an error is encountered.

After start-up, it remains possible to ask the command interpreter to process a preset list of commands from a file. This is done using the *do* command. This command requires a file-name argument; once it opens the file, the command interpreter simply reads commands as if they had been typed in by the user. When it encounters an error, it prints the corresponding error message and presents the following *interrupt* prompt:

iac: p to push /b to break / <cr> to continue:

If you enter *p*, the command interpreter will be called recursively, and you can enter any commands you wish (see below). If you enter *b*, the command interpreter will quit reading commands from the command file and return to the top level for further input. The notation *<cr>* stands for carriage return; if you just type a *return*, the command interpreter will continue trying to read commands from the command file, with possibly anomalous results. In general, it is best to break at this point and to quit the program and fix the *do* file.

Interrupting Processing

Sometimes when the program is running, you may see that it is doing something you had not intended or that it is continuing longer than you wanted. In this case, you can type *control-C* (hold down the key marked *control* and the *C* key at the same time). This causes the program to set a flag that is checked at the end of the current processing cycle. When the flag is found to be set, the program prints the interrupt prompt, giving the user the option to push, break, or continue. Again, you can continue processing by typing *return*; you can break and return to command level by entering *b*; or you can call the command interpreter recursively by entering *p*.

When you interrupt processing, the interrupt character (^C) will appear at a random place on the screen. If this is annoying, you may enter *display state* to the command interpreter to clear the screen and redisplay the state of the network.

Single Stepping

In order to examine the course of processing as it unfolds, the program offers the user the option of using the *single-step* mode. When this mode is on, the program interrupts itself after each processing step, displaying the interrupt prompt, just as if the user had typed an interrupt. Generally, the user will type *return* when ready to continue, but it remains possible to either break or push to a recursive command level.

The Recursive Command Level

When the command interpreter is called recursively, it displays the following prompt:

[N] iac:

Here [*N*] indicates the depth of the recursion (it is possible to embed recursive calls indefinitely, or at least until your computer runs out of stack space). In this mode you can enter commands as you normally would. To terminate this recursive call to the command interpreter, simply type *return*; this will return you to the interrupt prompt, from which the push, break, and continue options remain available.

Running Commands Outside the Program

Sometimes, while a simulation is in progress, it is useful to execute a command outside the simulation program. For example, you may wish to determine whether a particular file exists. To do this you can use the *run* command. Simply enter *run*, followed by the command, followed by any arguments to the command, followed by *end*. For example, on a PC, to see if the file *foo.log* exists, you would enter:

run dir foo.log end

The program will then pass the command

dir foo.log

to the MS-DOS command interpreter.
As another example, if you enter

run command end

the MS-DOS command interpreter will be invoked, and you can enter MS-DOS commands. To return to the simulation program at this point, just enter *exit*.

Quitting the Program

To quit the program, you enter *quit* to the command interpreter. Since you may have done various things that you would like to save before you quit, the program asks you to confirm your intention of quitting. If you enter *y* at this point, the program will terminate, closing all files that were open and clearing the screen. If you do not want to quit, type *return* or anything else.

Display Conventions

Displays are organized to pack lots of information into a small space. As a result, values of floating-point variables are usually given in a compact format in which only two or three character positions are used for each variable. The true value of the variable is first multiplied by a scale factor (specified in the template file) and then printed in the available space. For activations and other variables with absolute values less than 1.0, a scale factor of 100 is generally used, so that an activation of 0.01 displays as " 1" and 0.99 displays as " 99." Values that are negative are displayed, by default, in *standout* mode, which on most display devices is reverse video (lighted background). If you are running the programs on a display device that has no standout mode, you should turn off the use of standout by using the *display options standout* command described in the command description section. When standout mode is off (and when the screen image is saved in a file), negative numbers are displayed using minus signs where possible. Thus, if three character positions are available, −0.99 displays as "−99."

Special provisions have been made to deal with the problem of displaying variables that cannot fit the space available to them. Values that exceed the available space are printed as one or more "*" characters. When standout mode is off and the minus sign will not fit, the number is printed in an alphabetic code, where *a* to *j* stand for the digits 1 to 9 and *o* stands for 0; values that are still too large to fit are displayed as one or more *X*'s.

Making Graphs

Utility programs are provided for making simple graphs of output that has been stored in log files. The use of these utility programs is described in Appendix D. You may also wish to use other graphics software of your own, since the plotting program we provide has very coarse spatial resolution.

Command Descriptions

Here we describe all of the commands available in the **iac** program. We suggest you take a brief look at these descriptions of the commands and at the following discussion of the variables used in the program, to get a sense of how things are organized, and then proceed to the first exercise, where you will have a chance to learn about the commands by using them. (A list of all of the commands available in all programs is given in Appendix B.)

Note that we define a command as a sequence consisting of a command verb followed by one or more command specifiers. First the simple top-level commands are described: These are the commands that consist only of a command verb followed by arguments. Then the more general command verbs and the specific commands available using each verb are described in a nested fashion. We first describe the command verb and then give the specifiers for that verb; if there are subspecifiers, they are further nested under the specifiers.

The top-level commands in the program allow the user to run the simulation model and to carry out a variety of other actions. Most of these commands are common to all of the programs presented in this book.

clear

Clears the screen, leaving it blank.

cycle

Runs the program through *ncycles* processing cycles.

do

Prompts for a file name containing a list of commands to execute and then prompts for a count, which indicates the number of times to execute the entire list.

input

Prompts for unit names or numbers and then prompts for external input values. These may be any real number, though the usual values are +1 and −1. Entered values are placed in the external input vector for use during subsequent processing.

log

Prompts for a file name to store a log of the information displayed to the screen. If the file the user specifies already exists, the new log will be appended to the end of this file. Once a log file has been opened, each time the screen is updated, all the templates that are displayed are checked to see if they should be logged. All the information logged at one time is placed on a single line in the log file, separated by spaces. Items whose *dlevel* is less than or equal to the global *slevel* parameter are logged in the order they occur in the *.tem* file; all other items are skipped over. See Appendix D for a more detailed explanation. Logging can be turned off by calling *log* again and entering "−" instead of a file name; alternatively, a new file name will close the old log file and open a new one.

quit

Quits the program. Prompts for confirmation; to confirm, enter *y*.

reset

Resets the model to its initial state. The *cycleno* is set to 0, and the activations of all the units are set back to the resting level. The display is cleared and updated.

run

Passes a command out of the program for execution by the MS-DOS command interpreter. Prompts for a command, then for arguments. End the list of arguments by entering *end* or by simply typing an extra *return*.

test

Prompts for the name or number of a pattern to test the model with. If the pattern is successfully identified, the program uses it to set the values of the first *ninputs* elements of the external input vector; all additional units in the net are given an external input of 0. This command then resets the network and runs the program through *ncycles* of processing. The user may then run more cycles by using the *cycle* command.

This ends the simple commands. We now turn to the compound commands that require a general command followed by further specifiers. As in the program itself, we designate command words that require further specifiers with "/"; the user is not required to type these, however. We describe each general command in turn, giving each with a list of the further specifiers available.

disp/

Allows the user to display the current state of the network, to designate a specific template for display, or to set various display options, using the *opt* subcommand.

disp/ state

Clears the screen and displays the current state of the network, as specified by the global *dlevel* variable and by the attributes of the various templates found in the *.tem* file. All templates whose associated *dlevels* are less than or equal to the global *dlevel* are displayed.

disp/ <template>

Displays the designated template, regardless of its *dlevel*. All other template attributes are honored.

disp/ opt/

Allows the user to modify various display options using the subspecifiers that follow.

disp/ opt/ standout

Determines whether negative numbers are displayed to the screen in *standout* mode (reverse video, on most displays) or not. If standout mode is not available, you should set this option to 0;

otherwise, the program will attempt to use standout mode with negative numbers, and they will be indistinguishable from positive numbers.

disp/ opt/ <template>

Allows the user to change various options associated with the templates defined in the *.tem* file. After the user enters

disp opt <template>

the program prompts for an attribute of the template to change—either the display level, the number of digits of precision, or the scale factor associated with the template. The user then enters the name of the attribute (just the first character will do) followed by the new value to assign to this attribute.

exam/

Allows the user to examine and optionally set the value of one of the variables of the program. Use of this command is described under "Accessing Variables," later in this chapter. (This command is a synonym of *set/*.)

get/

Allows the user to get lists of things into the program, according to the option specified. The *get* commands either read a file or request a list of items terminated by *end* or a blank line.

get/ network

Allows the reading of a network specification from a file. The program prompts for the file name. The *get/ network* command is usually given in the *.str* file. A full description of the format of the network file is given in Appendix C.

get/ patterns

Allows the user to read in a set of input patterns from a file. Before it can be used, the variable *ninputs* must be defined. This variable tells the program how many input activation values to expect to find in each input pattern. The command prompts for a file name. Conventionally such files are given a *.pat* extension. A *.pat* file consists of a sequence of pattern specifications, typically one per line. The first entry in each pattern specification is a name for the pattern. Subsequent elements specify values to be used to specify inputs to each of the units in the network. Elements are separated by spaces. The first element determines the input to the first unit, the second element determines the input to the second unit, and so on. Elements may be floating-point numbers, or they may be a "+", "–", or "." character. A "+" indicates that the input to the corresponding unit is +1, a "–" indicates that the input is –1, and a "." indicates that the corresponding unit will receive a 0 input when the pattern is presented to the network. The number of patterns read is stored in the variable *npatterns*. The patterns are stored in an array called *ipattern*; each pattern is a separate row of this array. Thus, element 4 of pattern 2 is stored in *ipattern*[2][4].

get/ unames
: Allows the user to enter a list of names for the units in the model. Prompts for one name at a time; these are assigned sequentially to the units, starting with unit 0. The end of the list is indicated by typing *return* (a blank line) or entering the string *end.*

get/ weights
: Allows the reading of a file containing numerical weights, one for each connection in the network. Prompts for the name of a weight file. Weight files are expected to be in the same format as is used by the *save/ weights* command (see below).

save/
: Allows the user to save various kinds of information in files. There are two specifiers available in **iac**: *screen* and *weights.*

save/ screen
: Prompts for a file to store an image of the screen. If the file already exists, the screen image is appended to it. Standout mode is turned off since it cannot be "printed" in a file. The screen is displayed with standout mode off before the file name is requested; if you do not want to store what you see, simply type *return* instead of a file name, and no saving will take place.

save/ weights
: Prompts for a file to store the current values of the weights. If the file already exists, you are warned and asked if you wish to overwrite it. A file so created can later be read in again using the *get/ weights* command. Entries in the file are ordered as follows: For each unit, all the weights to that unit are stored in increasing order. After all the weights come the biases for each unit, if there are any biases. (There are no biases in **iac**.)

set/
: Allows the user to examine and optionally set the value of one of the variables of the program. Use of this command is described under "Accessing Variables," later in this chapter. (This command is a synonym of *exam/*.)

Variable Types

Variables are of several types: Some are strings of characters, some are integers, and some are floating-point numbers. Within each of these types, some are single-valued variables; others are vectors, or lists of variables with an index; and others are matrices, or two-dimensional arrays of variables with two indexes. In accessing single-valued variables, once the variable has been entered, the program will display the current value, and at this point the user may enter a new value or simply type *return* to return to the command level. With vector variables, an index must be given before a value is displayed for possible alteration; with array variables, two indexes are required.

Vector indexes may be given as numbers (with the first element being element 0) or, if the elements of a vector have names, these may be used as indexes. Thus, if unit 0 has been named *Fred*, the index 0 may be specified by entering *Fred*. Note that both units and patterns may have names. The program is set up so that unit names and pattern names are consulted; the program can get confused if the same names are used for both units and patterns.

Arrays are generally used for weights or other variables that are associated with the connections to a particular unit from another unit. The indexes are specified in receiver-sender order. Note that if the units have names, the names can be used here as well. Arrays are also used for patterns, with the first index specifying the pattern number and the second specifying the number of the element within the pattern.

All of the programs have a fairly large number of variables. These variables are organized into several different functional groups:

- *Mode.* These are variables whose values determine switchable characteristics of the model being implemented by the program. Generally, the programs can be thought of as implementing a "base" model and several variants. When the *mode* variables are all set to 0 (off), the base model is implemented. When one or more of these variables is set to 1 (on), other variants are in force.

- *Configuration.* These variables determine basic configurational properties of the network that is being used in a particular simulation run, and they are generally set in the *.net* file. Configuration variables should generally not be changed during a run but can be examined. Also included as a configuration variable is the list of names that have been assigned to the units in the network.

- *Environment.* These are variables associated with the test environment in which the program is run, that is, with the set of patterns that may be presented to the model for testing.

- *Parameter.* These variables are the parameters of the model, the ones that determine such things as the relative strength of excitation vs. inhibition, the decay rate, and so on.

- *State.* These are variables that are associated with the current state of the processing network, such as the activation values of the units.

- *Top-level variables.* These variables are the ones you need to change to control the activity of the simulation model itself. Also included at the top level are the weights associated with the connections and the bias terms (if any) associated with the units in the network.

Accessing Variables

Variables are accessed using the *set* and *exam* commands. The top-level variables are accessible directly, because they tend to be accessed most often in using the programs. These variables are accessed by typing

set <variable>

or

exam <variable>

Other variables are accessed through specifiers that correspond to the different variable types; the specifiers are *config, mode, env, param*, and *state.* They are accessed by typing

set <specifier> <variable>

or

exam <specifier> <variable>

Variable List

Here follows a list of all of the variables available in **iac**. First listed are all of variables directly accessible via the *set* and *exam* commands. These are followed by the variables that require specifiers, as indicated.

dlevel
: An integer variable that determines which templates will be displayed when the display is updated. All templates with *dlevel*s as specified in the *.tem* file that are less than or equal to this global *dlevel* parameter are updated.

ncycles
: An integer variable that specifies how many processing cycles are executed when the *cycle* command is entered.

seed
: The current value of the seed used by the random number generator. May be set by the user to equal any integer. Not used in **iac**; see Chapter 3 for a full discussion.

single
: A switch variable, normally set to 0, that makes the program run in single-step mode when set to 1.

slevel
: An integer variable that determines which templates will be logged in the log file when the screen is updated. Note that a template is logged only if (a) a log file has been opened with the *log* command, (b) the template's *dlevel* is less than or equal to the global *dlevel*, and (c) the template's *dlevel* is also less than the global *slevel.*

stepsize
: A string variable that controls the size of the processing steps taken by the program between screen updates. Allowed values are *cycle* and *ncycles*.

weight
: A floating-point matrix variable. The matrix contains the weights to each unit in the network from each unit. The user can examine the value of a particular weight by entering the command *set/ weight <to_index> <from_index>*, where *<to_index>* is the name or number of the receiving unit and *<from_index>* is the name or number of the sending unit.

config/ ninputs
: An integer variable specifying the number of input units for which external inputs will be specified in each *ipattern* read by the *get/ patterns* command.

config/ nunits
: An integer variable specifying the number of units in the network. This is generally declared in the *.net* file and should not be reset but may be examined.

config/ uname
: A vector of character strings read in by the *get/ unames* command. These strings are taken to be the names associated with the units in the network.

env/ ipattern
: A floating-point array variable containing the patterns read in by the last *get/ patterns* command. Note that the first index specifies the *pattern*, and the second index specifies the *element* within the pattern. Thus the command *exam/ ipattern 3 2* requests the program to print the value of element 2 in pattern 3. Note, the term *ipattern* is used rather than just *pattern* because in some programs there are two types of patterns, input patterns and target patterns. The *ipattern* variable is used to refer to the former type.

env/ maxpatterns
The maximum number of patterns that can be read into the network. This variable is automatically increased by the program if more patterns are encountered in the *.pat* file, but when large numbers of patterns are used, network initialization occurs more quickly if *maxpatterns* is set to a value a bit greater than the number of patterns that will be read in.

env/ npatterns
The number of patterns that have been read into the program by the last *get/ patterns* command.

env/ pname
A vector or list of character string variables specifying the names of the patterns read in by the *get/ patterns* command.

mode/ gb
This mode variable, when set to 1, causes the program to use Grossberg's updating function rather than the standard function taken from McClelland and Rumelhart (1981).

param/ alpha
A floating-point variable that scales the strength of the excitatory influences on units from other units.

param/ decay
A floating-point variable specifying the decay rate of unit's activations.

param/ estr
A floating-point variable that scales the strength of the external input to units in the network.

param/ gamma
A floating-point variable that scales the strength of the inhibitory influences on units from other units.

param/ max
A floating-point variable specifying the maximum activation of each unit.

param/ min
A floating-point variable specifying the minimum activation of each unit.

param/ rest
A floating-point variable specifying the resting activation of each unit.

state/ activation
A vector of floating-point variables specifying the activation values of the units in the network.

state/ cpname
A string variable specifying the name of the current pattern (if any) that was specified for testing via the *test* command.

state/ cycleno
The number of processing cycles elapsed since the last reset.

state/ excitation
A vector of floating-point values specifying the excitatory input to each unit, as computed during the most recent processing cycle.

state/ extinput
A vector of floating-point values specifying the external input to each unit, as determined by the last *input* command or by the last pattern specified for testing via the *test* command.

state/ inhibition
A vector of floating-point values specifying the inhibitory input to each unit, as computed during the most recent processing cycle.

state/ netinput
A vector of floating-point values specifying the net input to each unit, as computed during the most recent processing cycle.

state/ patno
The index of the current pattern (if any) that was specified for testing via the *test* command.

OVERVIEW OF EXERCISES

In this section we suggest several different exercises. Each will stretch your understanding of IAC networks in a different way. Ex. 2.1 focuses primarily on basic properties of IAC networks and their application to various problems in memory retrieval and reconstruction. Ex. 2.2 suggests experiments you can do to examine the effects of various parameter manipulations. Ex. 2.3 fosters the exploration of Grossberg's update rule as an alternative to the default update rule used in the **iac** program. Ex. 2.4 suggests that you develop your own task and network to use with the **iac** program.

If you want to cement a basic understanding of IAC networks, you should probably do several parts of Ex. 2.1, as well as Ex. 2.2. The first few parts of Ex. 2.1 also provide an easy tutorial example of the general use of the programs in this book. Answers to the questions in Exs. 2.1-2.3 are given in Appendix E.

Ex. 2.1. Retrieval and Generalization

Use the **iac** program to examine how the mechanisms of interactive activation and competition can be used to illustrate the following properties of human memory:

- Retrieval by name and by content.

- Retrieval with noisy cues.
- Assignment of plausible default values when stored information is incomplete.
- Spontaneous generalization over a set of familiar items.

The "data base" for this exercise is the Jets and Sharks data base shown in Figure 10 of *PDP:1* and reprinted here for convenience in Figure 1. You are to use the **iac** program in conjunction with this data base to run illustrative simulations of these basic properties of memory. In so doing, you will observe behaviors of the network that you will have to explain using the analysis of IAC networks presented earlier in the "Background" section.

The Jets and The Sharks

Name	Gang	Age	Edu	Mar	Occupation
Art	Jets	40's	J.H.	Sing.	Pusher
Al	Jets	30's	J.H.	Mar.	Burglar
Sam	Jets	20's	COL.	Sing.	Bookie
Clyde	Jets	40's	J.H.	Sing.	Bookie
Mike	Jets	30's	J.H.	Sing.	Bookie
Jim	Jets	20's	J.H.	Div.	Burglar
Greg	Jets	20's	H.S.	Mar.	Pusher
John	Jets	20's	J.H.	Mar.	Burglar
Doug	Jets	30's	H.S.	Sing.	Bookie
Lance	Jets	20's	J.H.	Mar.	Burglar
George	Jets	20's	J.H.	Div.	Burglar
Pete	Jets	20's	H.S.	Sing.	Bookie
Fred	Jets	20's	H.S.	Sing.	Pusher
Gene	Jets	20's	COL.	Sing.	Pusher
Ralph	Jets	30's	J.H.	Sing.	Pusher
Phil	Sharks	30's	COL.	Mar.	Pusher
Ike	Sharks	30's	J.H.	Sing.	Bookie
Nick	Sharks	30's	H.S.	Sing.	Pusher
Don	Sharks	30's	COL.	Mar.	Burglar
Ned	Sharks	30's	COL.	Mar.	Bookie
Karl	Sharks	40's	H.S.	Mar.	Bookie
Ken	Sharks	20's	H.S.	Sing.	Burglar
Earl	Sharks	40's	H.S.	Mar.	Burglar
Rick	Sharks	30's	H.S.	Div.	Burglar
Ol	Sharks	30's	COL.	Mar.	Pusher
Neal	Sharks	30's	H.S.	Sing.	Bookie
Dave	Sharks	30's	H.S.	Div.	Pusher

FIGURE 1. Characteristics of a number of individuals belonging to two gangs, the Jets and the Sharks. (From "Retrieving General and Specific Knowledge From Stored Knowledge of Specifics" by J. L. McClelland, 1981, *Proceedings of the Third Annual Conference of the Cognitive Science Society*. Copyright 1981 by J. L. McClelland. Reprinted by permission.)

Starting up. Before running this exercise, you must first set up an *iac* directory as described in Appendix A. Change your working directory to this directory, and enter

iac jets.tem jets.str

This causes the program to begin running using the template information stored in the *jets.tem* file, with the start-up commands contained in the *jets.str* file. The latter file contains the command

get net jets.net

This command causes the program to set up a network containing 68 units. The units are grouped into seven pools: a pool of *name* units, a pool of *gang* units, a pool of *age* units, a pool of *education* units, a pool of *marital status* units, a pool of *occupation* units, and a pool of *instance* units. The *name* pool contains a unit for the name of each person; the *gang* pool contains a unit for each of the gangs the people are members of (Jets and Sharks); the *age* pool contains a unit for each age range; and so on. The pool of *instance* units contains a unit for each individual in the set.

The units in the first six pools can be called *visible* units, since all are assumed to be accessible from outside the network. Those in the gang, age, education, marital status, and occupation pools can also be called property units. The instance units are assumed to be inaccessible, so they can be called *hidden* units.

Each unit has an inhibitory connection to every other unit in the same pool. In addition, there are two-way excitatory connections between each instance unit and the units for its properties, as illustrated in Figure 2 (Figure 11 from *PDP:1*). Note that the figure is incomplete, in that only some of the name and instance units are shown. The *jets.str* file provides names for each of the units, using the *get/ names* command. These are given only for the convenience of the user, of course; all actual computation in the network occurs only by way of the connections.

The *jets.str* file also sets the values of the parameters of the model. These values are

decay = 0.1
alpha = 0.1
gamma = 0.1
estr = 0.4
max = 1.0
min = −0.2
rest = −0.1

These are all set using the command form:

set param <*name*> <*value*>

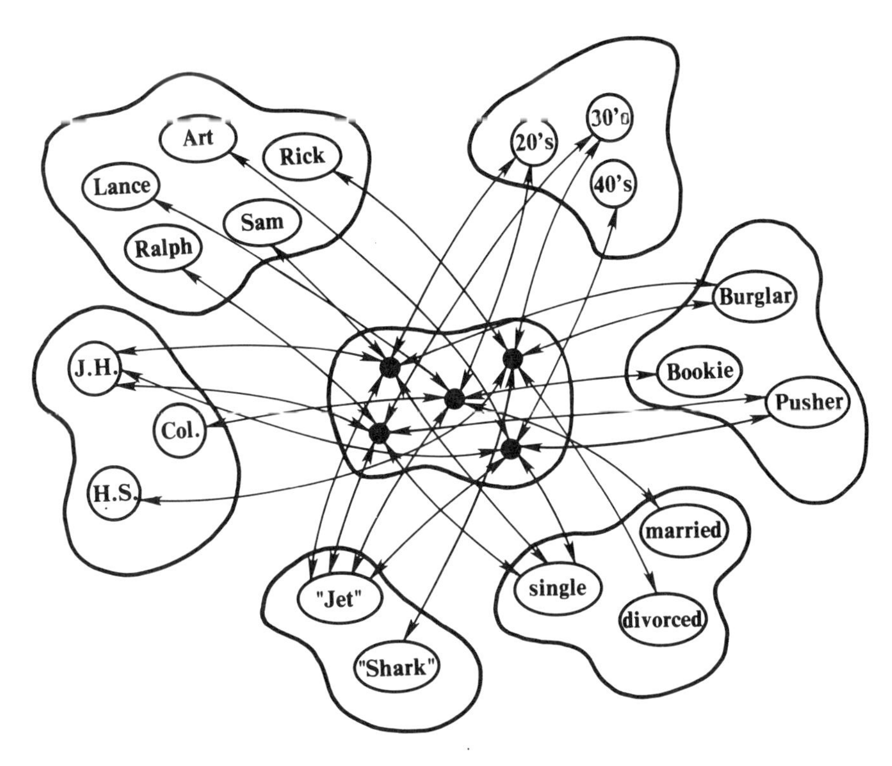

FIGURE 2. The units and connections for some of the individuals in Figure 1. (Two slight errors in the connections depicted in the original of this figure have been corrected in this version.) (From "Retrieving General and Specific Knowledge From Stored Knowledge of Specifics" by J. L. McClelland, 1981, *Proceedings of the Third Annual Conference of the Cognitive Science Society*. Copyright 1981 by J. L. McClelland. Reprinted by permission.)

After the program has read the information contained in the *jets.tem* and *jets.str* files, it produces the display shown in Figure 3. The display shows the names of all of the units, each flanked with a number on the left and a number on the right. The number on the left indicates the external input to the unit; since no input has been specified, these numbers are now all 0. The number on the right indicates the unit's current activation value, which at this point is equal to its resting activation. These numbers are to be read as *hundredths*. Thus this initial display indicates that all of the units have a resting activation level of −0.10.

Units are organized into columns, with the property units in the first column, the name units for the Jets in the second column, the name units for the Sharks in the third column, and the instance units for the Jets and the Sharks in the fourth and fifth columns, respectively. Note that the names of the instance units begin with "_"; the name unit for Lance is called *Lance* but the instance unit for Lance is called *_Lance*. On the far right of the display is the current cycle number, which is initialized to 0.

```
iac:
disp/  exam/  get/  save/  set/  clear  cycle  do  input  log  quit  reset  run
test

 0 Jets     10   0 Art     10   0 Phil   10   0 _Art     10   0 _Phil  10  cycle   0
 0 Sharks   10   0 Al      10   0 Ike    10   0 _Al      10   0 _Ike   10
                 0 Sam     10   0 Nick   10   0 _Sam     10   0 _Nick  10
 0 in20s    10   0 Clyde   10   0 Don    10   0 _Clyde   10   0 _Don   10
 0 in30s    10   0 Mike    10   0 Ned    10   0 _Mike    10   0 _Ned   10
 0 in40s    10   0 Jim     10   0 Karl   10   0 _Jim     10   0 _Karl  10
                 0 Greg    10   0 Ken    10   0 _Greg    10   0 _Ken   10
 0 JH       10   0 John    10   0 Earl   10   0 _John    10   0 _Earl  10
 0 HS       10   0 Doug    10   0 Rick   10   0 _Doug    10   0 _Rick  10
 0 College  10   0 Lance   10   0 Ol     10   0 _Lance   10   0 _Ol    10
                 0 George  10   0 Neal   10   0 _George  10   0 _Neal  10
 0 Single   10   0 Pete    10   0 Dave   10   0 _Pete    10   0 _Dave  10
 0 Married  10   0 Fred    10                 0 _Fred    10
 0 Divorce  10   0 Gene    10                 0 _Gene    10
                 0 Ralph   10                 0 _Ralph   10
 0 Pusher   10
 0 Burglar  10
 0 Bookie   10
```

FIGURE 3. The initial display produced by the **iac** program for Exercise 1.

At the top of the initial display is the program prompt, **iac:**, as well as a menu of all of the command words that can be entered by the user.

Since everything is set up for you, you are now ready to do each of the separate parts of the exercise. Each part is accomplished by using the interactive activation and competition process to do pattern completion, given some probe that is presented to the network. For example, to retrieve an individual's properties from his name, you simply provide external input to his name unit, then allow the IAC network to propagate activation first to the name unit, then from there to the instance units, and from there to the units for the properties of the instance.

Retrieving an individual from his name. To illustrate retrieval of the properties of an individual from his name, we will use Ken as our example. To specify external input to the name unit for Ken, you use the *input* command. When you type this command to the **iac:** prompt, the program asks if you want to clear all previous inputs (there are none, so you can enter *y* or *n*). Then the command prompts for a unit name or number, followed by an external input value for that unit. In this case you simply enter *Ken* as the name and 1 as the input value.

The program will accept as many name-value pairs as you wish to enter; to stop entering pairs just enter *end* or type *return*. When you've done this, the screen will be updated to indicate the external inputs you have specified. Note that external input of 1.00 displays as "**"; so if you have provided external input to *Ken* with value 1, there will be two stars instead of a 0 next to his name. After this update, you will get back the **iac:** prompt, and you are ready to run your first simulation.

To start processing, enter the *cycle* command. This causes the program to run for 10 cycles. As it runs each cycle, it will update the screen, but this will generally occur rather quickly. If you'd like to watch the activation process unfold one cycle at a time, you can place the program in single-step mode by entering

set single 1

Having turned on single mode, you can continue to run more cycles if you wish. Alternatively, you may want to start over. To do that, enter *reset*, then enter *cycle* again. Now the screen will be updated after each cycle and processing will pause with the prompt:

p to push /b to break / <cr> to continue:

When you are ready to let the program run the next cycle just type *return*. In this way you can step through the cycles, one at a time, and examine at your leisure what happened on each cycle. Processing will again terminate at the end of 10 cycles. If you would like to allow processing to go on for longer, you can simply set *ncycles* to a larger number, say 100, by entering

set ncycles 100

As you will observe, activations continue to change for many cycles of processing. Things slow down gradually, so that after a while not much seems to be happening on each trial. Eventually things just about stop changing. Once you've run about 100 cycles, you'll have about reached asymptotic activation values. (You may, of course, find it convenient to set single back to 0 at some point in this process.)

A picture of the screen after 100 cycles is shown in Figure 4. At this point, you can check to see that the model has indeed retrieved the pattern for Ken correctly. There are also several other things going on that are worth understanding. Try to answer all of the following questions (you'll have to refer to the properties of the individuals, as given in Figure 1).

Q.2.1.1. Why are some of Ken's properties more strongly activated than others?

Q.2.1.2. None of the other name units were activated, yet several other instance units are active (i.e., their activation is greater than 0). Explain this difference.

Q.2.1.3. Why are some of the active instance units more active than others?

Q.2.1.4. The *in30s* unit is receiving almost as much excitation as the *in20s* unit. Why is the latter so much more active than the former?

```
iac:
disp/  exam/  get/  save/  set/  clear  cycle  do  input  log  quit  reset  run
test

0 Jets     12  0 Art     14  0 Phil   14  0 _Art     14  0 _Phil  14  cycle 100
0 Sharks   50  0 Al      14  0 Ike    14  0 _Al      14  0 _Ike   12
               0 Sam     14  0 Nick   13  0 _Sam     12  0 _Nick  23
0 in20s    37  0 Clyde   14  0 Don    14  0 _Clyde   14  0 _Don   12
0 in30s     0  0 Mike    14  0 Ned    14  0 _Mike    14  0 _Ned   14
0 in40s    12  0 Jim     14  0 Karl   14  0 _Jim     13  0 _Karl  12
               0 Greg    14 ** Ken    80  0 _Greg    12  0 _Ken   68
0 JH       13  0 John    14  0 Earl   14  0 _John    13  0 _Earl   3
0 HS       52  0 Doug    14  0 Rick   14  0 _Doug    12  0 _Rick   3
0 College  13  0 Lance   14  0 Ol     14  0 _Lance   13  0 _Ol    14
               0 George  14  0 Neal   13  0 _George  13  0 _Neal  23
0 Single   50  0 Pete    14  0 Dave   14  0 _Pete     3  0 _Dave  12
0 Married  13  0 Fred    14               0 _Fred     3
0 Divorce  13  0 Gene    14               0 _Gene    12
               0 Ralph   14               0 _Ralph   14
0 Pusher   11
0 Burglar  37
0 Bookie   11
```

FIGURE 4. The display screen after 100 cycles with external input to the name unit for Ken.

If you can answer all of these questions correctly, you understand the interactive activation and competition process rather well.

Retrieval from a partial description. Next, we will use the **iac** program to illustrate how it can retrieve an instance from a partial description of its properties. We will continue to use Ken, who, as it happens, can be uniquely described by two properties, *Shark* and *in20s.* To do this, enter the *input* command, enter *y* to the question about clearing the previous inputs, and then specify external inputs of 1 for *Shark* and *in20s.* After getting back to the **iac:** prompt, enter *reset* and then enter *cycle.* Run a total of 100 cycles again, and take a look at the state of the network.

Q.2.1.5. Describe the differences between this state and the state at the end of the previous run. What are the main differences?

Q.2.1.6. Explain why the occupation units show partial activations of units other than Ken's occupation, which is Burglar. Take the explanation as far as you can, contrasting the current case with the previous case as much as possible.

In comparing this case with the previous one, it may be useful for you to make a copy of the state of the screen for future reference. To do so, just enter

save screen

and give a file name as requested. Note that negative numbers are saved with minus signs instead of in reverse video, so that they can easily be printed.

Graceful degradation. What is the effect of erroneous information in the probe supplied to an IAC network? Here we examine how errors influence the ability of the network to retrieve an individual's name from a description of his properties. To begin to explore this issue, run the model twice: once activating all of Ken's properties other than his name via the input command and once with the same external input, but activating *JH* instead of *HS*. Note that for the second case, you can retain the old input specification, and then just set the input to *HS* to 0 and the input to *JH* to 1 before retesting.

Q.2.1.7. How well does the model do with the "noisy" version of Ken compared to the correct version of Ken? Would it do this well with all noisy versions of individuals? Test with at least one other individual, and explain your results.

Default assignment. Sometimes we do not know something about an individual; for example, we may never have been exposed to the fact that Lance is a Burglar. Yet we are able to give plausible guesses about such missing information. The **iac** program can do this too. We illustrate by using the *set/ weights* command to set the weights between the instance unit for Lance and the property unit for Burglar to 0. First, run 100 cycles, providing external input of 1 to *Lance*, to see what happens before we delete the connections. Then, remove these connections as follows. To the **iac:** prompt type

set weight Burglar _Lance 0
set weight _Lance Burglar 0

Then reset the network, and run 100 cycles again.

Q.2.1.8. Describe how the model was able to fill in what in this instance turns out to be the correct occupation for Lance. Also, explain why the model tends to activate the *Divorced* unit as well as the *Married* unit.

Spontaneous generalization. Now we consider the network's ability to retrieve appropriate generalizations over sets of individuals—that is, its ability to answer questions like "What are Jets like?" or "What are people who are in their 20s and have only a junior high education like?" Be sure to reinstall the connections between *Burglar* and *_Lance* (set them back to 1). Once you've done that, you can ask the model to generalize about the Jets by providing external input to the Jets unit alone, then cycling for 100 cycles; you can ask it to generalize about the people in their 20s with a junior high education by providing external input to the *in20s* and *JH* units.

Q.2.1.9. Describe the strengths and weaknesses of the IAC model as a model of retrieval and generalization. How does it compare with other models you are familiar with? What properties do you like, and what properties do you dislike? Are there any general principles you can state about what the model is doing that are useful in gaining an understanding of its behavior?

Ex. 2.2. Effects of Changes in Parameter Values

In this exercise, we will examine the effects of variations of the parameters *estr*, *alpha*, *gamma*, and *decay* on the behavior of the **iac** program.

Increasing and decreasing the values of the strength parameters. Explore the effects of adjusting all of these parameters proportionally, using the partial description of Ken as probe (that is, providing external input to *Shark* and *in20s*).

Q.2.2.1. What effects do you observe from decreasing the values of *estr*, *alpha*, *gamma*, and *decay* by a factor of 2? What happens if you set them to twice their original values? See if you can explain what is happening here.

Hint. For this exercise, you should first consider the asymptotic activations of units—what do you expect based on the discussion in the "Background" section? If you wish to follow the time course of activation, you can use the *log* command to store a record of processing in a file, then you can use the **colex** and **plot** utilities to make graphs of activations vs. cycle numbers for selected units. A discussion of how to do this is given in Appendix D.

Relative strength of excitation and inhibition. Return all the parameters to their original values, then explore the effects of varying the value of *gamma* above and below 0.1, again providing external input to the *Sharks* and *in20s* units. Also examine the effects on the completion of Lance's properties from external input to his name, with and without the connections between the instance unit for Lance and the property unit for Burglar.

Q.2.2.2. Describe the effects of these manipulations and try to characterize their influence on the model's adequacy as a retrieval mechanism.

Ex. 2.3. Grossberg Variations

Explore the effects of using Grossberg's update rule rather than the default rule used in the IAC model. This requires you to use the command

set mode gb 1

Now redo one or two of the simulations from Exercise 1.

Q.2.3.1. What happens when you repeat some of the simulations suggested in Exercise 1 with *gb* mode on? Can these effects be compensated for by adjusting the strengths of any of the parameters? If so, explain why. Do any subtle differences remain, even after compensatory adjustments? If so, describe them.

Hints. In considering the issue of compensation, you should consider the difference in the way the two update rules handle inhibition and the differential role played by the minimum activation in each update rule.

Ex. 2.4. Construct Your Own IAC Network

Construct a task that seems appropriate for an IAC network, along with a knowledge base (in the form of a *.net* file), and explore how well the network does in performing your task. You may find it useful to set up *.str*, *.tem*, and *.loo* files, in addition to the *.net* file. (Detailed specifications for the *.net*, *.tem*, and *.loo* files are given in Appendix C.)

Q.2.4.1. Describe your task, your knowledge base, and the experiments you run on it. Discuss the adequacy of the IAC model to do the task you have set it.

Hints. You might bear in mind if you undertake this exercise that you can specify virtually *any* architecture you want in an IAC network, including architectures involving several layers of units. You might also want to consider the fact that such networks can be used in low-level perceptual tasks, in perceptual mechanisms that involve an interaction of stored knowledge with bottom-up information, as in the interactive activation model of word perception, in memory tasks, and in many other kinds of tasks. Use your imagination, and you may discover an interesting new application of IAC networks.

CHAPTER 3

Constraint Satisfaction in PDP Systems

In the previous chapter we showed how PDP networks could be used for content-addressable memory retrieval, for prototype generation, for plausibly making default assignments for missing variables, and for spontaneously generalizing to novel inputs. In fact, these characteristics are reflections of a far more general process that many PDP models are capable of—namely, finding near-optimal solutions to problems with a large set of simultaneous constraints. This chapter introduces this constraint satisfaction process more generally and discusses three different specific models for solving such problems. The specific models are the *schema model,* described in *PDP:14*; the *Boltzmann machine,* described in *PDP:7*; and *harmony theory,* described in *PDP:6*. These models are embodied in the **cs** (constraint satisfaction) program. We begin with a general discussion of constraint satisfaction and some general results. We then turn to the schema model. We describe the general characteristics of the schema model, show how it can be accessed from **cs**, and offer a number of examples of it in operation. This is followed in turn by detailed discussions of the Boltzmann machine and harmony theory.

BACKGROUND

Consider a problem whose solution involves the simultaneous satisfaction of a very large number of constraints. To make the problem more difficult, suppose that there may be no perfect solution in which all of the constraints are completely satisfied. In such a case, the solution would involve the satisfaction of as many constraints as possible. Finally, imagine that some constraints may be more important than others. In particular, suppose that each constraint has an importance value associated with it and that the solution to the problem involves the simultaneous satisfaction of as many of the most important of these constraints as possible. In general, this is a very difficult problem. It is what Minsky and Papert (1969) have called the *best match* problem. It is a problem that is central to much of cognitive science. It also happens to be one of the kinds of problems that PDP systems solve in a very natural way. Many of the chapters in the two PDP volumes pointed to the importance of this problem and to the kinds of solutions offered by PDP systems.

To our knowledge, Hinton was the first to sketch the basic idea for using parallel networks to solve constraint satisfaction problems (Hinton, 1977). Basically, such problems are translated into the language of PDP by assuming that each unit represents a hypothesis and each connection a constraint among hypotheses. Thus, for example, if whenever hypothesis A is true, hypothesis B is usually true, we would have a positive connection from unit A to unit B. If, on the other hand, hypothesis A provides evidence against hypothesis B, we would have a negative connection from unit A to unit B.

PDP constraint networks are designed to deal with *weak constraints* (Blake, 1983), that is, with situations in which constraints constitute a set of desiderata that *ought* to be satisfied rather than a set of *hard* constraints that *must* be satisfied. The goal is to find a solution in which as many of the most important constraints are satisfied as possible. The importance of the constraint is reflected by the strength of the connection representing that constraint. If the constraint is very important, the weights are large. Less important constraints involve smaller weights. In addition, units may receive external input. We can think of the external input as providing direct evidence for certain hypotheses. Sometimes we say the input "clamps" a unit. This means that, in the solution, this particular unit *must be on* if the input is positive or *must be off* if the input is negative. Other times the input is not clamped but is graded. In this case, the input behaves as simply another weak constraint. Finally, different hypotheses may have different a priori probabilities. An appropriate solution to a constraint satisfaction problem must be able to reflect such prior information as well. This is done in PDP systems by assuming that each unit has a *bias,* which acts to turn the unit on in the absence of other evidence. If a particular unit has a positive bias, then it is better to have the unit on; if it has a negative bias, there is a preference for it to be turned off.

We can now cast the constraint satisfaction problem described above in the following way. Let *goodness of fit* be the degree to which the desired constraints are satisfied. Thus, goodness of fit (or more simply *goodness*) depends on three things. First, it depends on the extent to which each unit satisfies the constraints imposed upon it by other units. Thus, if a connection between two units is positive, we say that the constraint is satisfied to the degree that both units are turned on. If the connection is negative, we can say that the constraint is violated to the degree that both units are turned on. A simple way of expressing this is to let the product of the activation of two units times the weight connecting them be the degree to which the constraint is satisfied. That is, for units i and j we let the product $w_{ij} a_i a_j$ represent the degree to which the pairwise constraint between those two hypotheses is satisfied. Note that for positive weights the more the two units are on, the better the constraint is satisfied; whereas for negative weights the more the two units are on, the less the constraint is satisfied. Second, the a priori strength of the hypothesis is captured by adding the bias to the goodness measure. Finally, the goodness of fit for a hypothesis when direct evidence is available is given by the product of the input value times the activation value of the unit. The bigger this product, the better the system is satisfying this external constraint. Given this, we can now characterize mathematically the degree to which a particular unit is satisfying all of the constraints impinging on it. Thus the overall degree to which the state of a particular unit, say unit i, contributes to the overall goodness of fit can be obtained by adding up the degree to which the unit satisfies all of the constraints in which it is involved, from all three sources. Thus, we can define the goodness of fit of unit i to be

$$goodness_i = \sum_j w_{ij} a_i a_j + input_i a_i + bias_i a_i . \tag{1}$$

This, of course, is just the sum of all of the individual constraints in which the corresponding hypothesis participates. It is not the individual hypothesis, however, that is the problem in constraint satisfaction problems. In these cases, we are concerned with the degree to which the entire pattern of values assigned to all of the units are consistent with the entire body of constraints. This overall goodness of fit is the function we want to maximize. We can define our overall goodness of fit as the sum of the individual goodnesses. In this case we get

$$goodness = \sum w_{ij} a_i a_j + \sum input_i a_i + \sum_i bias_i a_i . \tag{2}$$

We have solved the problem when we have found a set of activation values that maximizes this function. It should be noted that since we want to have the activation values of the units represent the degree to which a particular hypothesis is satisfied, we want our activation values to range between a minimum and maximum value—in which the maximum value is understood to mean that the hypothesis should be accepted and the minimum value means that it should be rejected. Intermediate values correspond to intermediate states of certainty.

We have now reduced the constraint satisfaction problem to the problem of maximizing the goodness function given above. There are many methods of finding the maxima of functions. Importantly, there is one method that is naturally and simply implemented in a class of PDP networks. One restriction on this class of networks is the restriction that the

Note: In Equation 2, each pair of units contributes only one time to the first summation (w_{ij} is assumed equal to w_{ji}). The total goodness is not the sum of the goodnesses of the individual units. It is best to think of the goodness of unit i as the set of terms in the total goodness to which the activation of unit i contributes.

weights in the network be symmetric: that is, the condition that $w_{ij} = w_{ji}$. Under these conditions it is easy to see how a PDP network naturally sets activation values so as to maximize the goodness function stated above. To see this, first notice that the goodness of a particular unit, $goodness_i$, can be written as the product of its current net input times its activation value. That is,

$$goodness_i = net_i a_i \tag{3}$$

where, as usual for PDP networks, net_i is defined as

$$net_i = \sum_j w_{ij} a_j + input_i + bias_i. \tag{4}$$

Thus, the net input into a unit provides the unit with information as to its contribution to the goodness of the entire solution. Consider any particular unit in the network. That unit can always behave so as to increase its contribution to the overall goodness of fit if, whenever its net input is positive, the unit moves its activation toward its maximum activation value, and whenever its net input is negative, it moves its activation toward its minimum value. Moreover, since the global goodness of fit is simply the sum of the individual goodnesses, a whole network of units behaving in such a way will always increase the global goodness measure. These observations were made by Hopfield (1982). We will return to Hopfield's important contribution to this analysis again in our discussion of Boltzmann machines and harmony theory.

It might be noted that there is a slight problem here. Consider the case in which two units are *simultaneously* evaluating their net inputs. Suppose that both units are off and that there is a large negative weight between them; suppose further that each unit has a small positive net input. In this case, both units may turn on, but since they are connected by a negative connection, as soon as they are both on the overall goodness may decline. In this case, the next time these units get a chance to update they will both go off and this cycle can continue. There are basically two solutions to this. The standard solution is not to allow more than one unit to update at a time. In this case, one or the other of the units will come on and prevent the other from coming on. This is the case of so-called *asynchronous* update. The other solution is to use a *synchronous* update rule but to have units increase their activation values very slowly so they can "feel" each other coming on and achieve an appropriate balance.

In practice, goodness values generally do not increase indefinitely. Since units can reach maximal or minimal values of activation, they cannot continue to increase their activation values after some point so they cannot continue to increase the overall goodness of the state. Rather, they increase it until they reach their own maximum or minimum activation values. Thereafter, each unit behaves so as to never decrease the overall goodness. In this way, the global goodness measure continues to increase until all units achieve their maximally extreme value or until their net input becomes exactly 0. When this is achieved, the system will stop changing and will have found a maximum in the goodness function and therefore a solution to our constraint satisfaction problem. When it reaches this peak in the goodness function, the goodness can no longer change and the network is said to have reached a *stable state;* we say it has *settled* or *relaxed* to a solution. Strictly speaking, this solution state can be guaranteed only to be a *local* rather than a *global* maximum in the goodness function. That is, this is a *hill-climbing* procedure that simply ensures that the system will find a peak in the goodness function, not that it will find the highest peak. The problem of local maxima is difficult. We address it at length in a later section. Suffice it to say, that different PDP systems differ in the difficulty they have with this problem.

The development thus far applies to all three of the models under discussion in this chapter. It can also be noted that if the weight matrix in an IAC network is symmetric, it too is an example of a constraint satisfaction system. Clearly, there is a close relation between constraint satisfaction systems and content-addressable memories.

We turn, at this point, to a discussion of the specific models and some examples with each. We begin with the schema model of *PDP:14*.

THE SCHEMA MODEL

The schema model is one of the simplest of the constraint satisfaction models, but, nevertheless, it offers useful insights into the operation of all of the constraint satisfaction models. In *PDP:2* we described a set of characteristics required to specify any model's particular features. The three models under discussion differ from one another primarily as to whether the units behave deterministically or stochastically (probabilistically), whether the units take on a continuum of values or only binary values, and by the allowable set of connections among the units. The schema model is deterministic; its units can take on any value between 0 and 1. The connection matrix is symmetric and the units may not connect to themselves (i.e., $w_{ij} = w_{ji}$ and $w_{ii} = 0$). Update in the schema model is asynchronous. That is, units are chosen to be updated sequentially in random order. When chosen, the net input to the unit is computed and the activation of the unit is modified. The logic of the hill-climbing method implies that whenever the net input (net_i) is positive we must increase the activation value of the unit, and when it is negative we must decrease the activation value. Thus we use the following simple update rule:

$$a_i(t+1) = a_i(t) + net_i(1 - a_i(t)) \tag{5}$$

when net_i is greater than 0, and

$$a_i(t+1) = a_i(t) + net_i a_i(t) \tag{6}$$

when net_i is less than 0. Note that in this second case, since net_i is negative and a_i is positive, we are decreasing the activation of the unit. This rule has two virtues: it conforms to the requirements of our goodness function and it naturally constrains the activations between 0 and 1.

As usual in these models, the net input comes from three sources: a unit's neighbors, its bias, and its external inputs. These sources are added. Thus, we have

$$net_i = istr\,(\sum_j w_{ij} a_j + bias_i) + estr\,(input_i). \tag{7}$$

Here the constants *istr* and *estr* are parameters that allow the relative contributions of the input from external sources and that from internal sources to be readily manipulated.

IMPLEMENTATION

The **cs** program implementing the schema model is much like **iac** in structure. It differs in that it does *asynchronous* updates using a slightly different activation rule. Like **iac**, **cs** consists of essentially two routines: (a) an update routine called *rupdate* (for random update), which selects units at random and computes their net inputs and then their new activation values, and (b) a control routine, *cycle*, which calls *rupdate* in a loop for the specified number of cycles while displaying the results on the screen. Thus, *cycle* is as follows:

```
cycle() {
  for(i = 0; i < ncycles; i++) {
    cycleno++;
    rupdate();
    update_display();
  }
}
```

Thus, each time *cycle* is called, the system calls *rupdate* and then displays the results of the computation. The *rupdate* routine itself does all of the work. It randomly selects a unit, computes its net input, and assigns the new activation value to the unit. It does this *nupdates* times. Typically, *nupdates* is set equal to *nunits*, so a single call to *rupdate*, on average, updates each unit once:

```
rupdate() {

  for (updateno = 0; updateno < nupdates; updateno++) {
    i = randint(0, nunits - 1);
    netinput = 0;
    for(j = 0; j < nunits; j++) {
      netinput += activation[j]*weight[i][j];
    }
    netinput += bias[i];
    netinput *= istr;
    netinput += estrength*input[i];

    if (netinput > 0)
      activation[i] += netinput*(1-activation[i]);
    else
      activation[i] += netinput*activation[i];
  }
}
```

RUNNING THE PROGRAM

The basic structure of **cs** and the mechanics of interacting with it are identical to those of **iac**. The **cs** program, like all of our programs, requires a template (*.tem*) file that specifies what is displayed on the screen and a start-up (*.str*) file that initializes the program with the parameters of the particular program under consideration. It also requires a *.net* file specifying the particular network under consideration, and may use a *.wts* file to specify particular values for the weights. It also allows for a *.pat* file for specifying a set of patterns that can be presented to the network. Once you are in the appropriate directory, the program is accessed by entering the command:

cs xxx.tem xxx.str

where *xxx* is the name of the particular example you are running.

The normal sequence for running the model involves applying external inputs to some subset of the units by use of the *input* command and using the *cycle* command to cause the network to cycle until it finds a goodness maximum. Typically, the value of the goodness is displayed after each cycle, and the system will cycle *ncycles* times and then stop. If the system has not yet reached a stable state, it can be continued from where it left off if the user simply enters *cycle* again. The system can be interrupted at any time by typing ^C (control-C).

Two commands are available for restarting the system. Both commands set *cycleno* back to 0, and both reinitialize the activations of all of the units.

However, one of these commands, *newstart*, causes the program to follow a new random updating sequence when next the *cycle* command is given, whereas the other command, *reset*, causes the program to repeat the same updating sequence used in the previous run. Alternatively, the user can specify a particular value for the random seed and enter it manually via the *set/ seed* command; when *reset* is next called, this value of the seed will be used, producing results identical to those produced on other runs begun with this same *seed* in force.

The **cs** program implements both the Boltzmann model and harmony theory in addition to the schema model. In this section we will introduce those aspects of **cs** relevant to all three models, even though some of these will not be explained until later in the chapter.

New or Altered Commands

newstart

Generates a new random seed for the random number generator and then issues a reset command. The effect is to cause the net to follow a new random sequence of updates when *cycle* is subsequently entered.

reset

Resets the network back to its initial state. In *clamp* mode, units with positive external input are clamped *on* and units with negative external inputs are clamped *off*. All others have their activations set to 0. The cycle number is reset to 0, and the random number generator is seeded with the value of the *seed* variable. Unless the *seed* has been changed, either by the *set/ seed* command or by calling *reset* via *newstart*, this means that the program will go on to repeat the same random sequence that was generated after the last *reset*.

get/ annealing

Allows the user to specify an annealing schedule for use in *boltzmann* or *harmony* mode (these modes are discussed later in the chapter). It begins by prompting for an initial temperature. The annealing schedule begins at this temperature. It then prompts for a sequence of time-temperature pairs. A carriage return or the string *end* given in response to the prompt will terminate the entry of the schedule. The system linearly interpolates from the initial temperature so that at the time (measured in number of cycles) specified in the first pair, the temperature will reach the value specified for that time. It then linearly interpolates from that temperature to the next temperature at the next time. This continues until the final time is reached. Thereafter the temperature remains constant at the final value.

Variables

The following variables are new or somewhat different in the **cs** program. They are accessed by way of the *set* and *exam* commands as in **iac**.

bias

A vector that contains the values of the biases for each of the units.

nupdates

Determines the number of updates per cycle. Generally, it is initialized in the *.net* file to be equal to *nunits*, so that each unit will be updated once per cycle, on the average.

seed

The current value of the seed used for reinitializing the random number generator. The *seed* is set to a random starting value when the program is first called, and this value is used to initialize the random number generator. Calls to *reset* cause the random number generator to be reinitialized to the current value of *seed*, and calls to *newstart* cause *seed* itself to be set to a new random value *before* resetting. The *seed* may be set to any desired integer value using the *set/ seed* command. Unless manually changed, the value of *seed* will be the value used last time the random number generator was reinitialized and it can be used again later to repeat the same sequence.

sigma

A vector that contains the values of the importance parameters associated with knowledge atoms in *harmony theory*.

stepsize

Determines exactly when information about the state of the program is displayed to the screen. If *stepsize* is set to *cycle* (the default value), the information will be displayed on the screen after every cycle. If the *stepsize* is set to *update*, information will be displayed on the screen after every time a unit is updated. If *single* is set to 1, the program will pause after every screen update.

mode/ boltzmann

If *boltzmann* is set to 1, the program will behave as a Boltzmann machine. If it is set to 0 it will act as the schema model. The default value is 0.

mode/ clamp

If *clamp* is set to 1, positive external inputs supplied via the *input* and *test* commands cause the units receiving them to come on and stay on, and negative inputs cause the units to go off and stay off. Zero inputs have no effect on the units. If *clamp* is set to 0, external inputs are graded and act as additional weak constraints on their units; they are simply added into the net input of the unit. The default is that *clamp* is set to 0.

mode/ harmony

If *harmony* is set to 1, the program will behave as a *harmonium*, as defined in *PDP:6*. The default is that this is set to 0.

param/ istr

Scales the magnitude of internal input to each unit.

param/ kappa

This parameter is relevant in *harmony* mode. It is basically a global threshold that indicates the fraction of the inputs to a knowledge atom that must "agree with" that knowledge atom before that atom will tend to come on. See the discussion of harmony theory for details.

state/ cuname

The name of the current unit (the one most recently updated).

state/ goodness

Contains the current value of the goodness for the entire network.

state/ temperature

A variable relevant to the Boltzmann machine and harmony theory, as will be explained in those sections. It is normally adjusted in accordance with an annealing schedule.

state/ unitno

The number of the unit last updated.

state/ updateno

Tells which update is currently being done, counting from the beginning of the current cycle.

OVERVIEW OF EXERCISES

We offer four exercises to try with the schema model. In Ex. 3.1, we give you the chance to explore the basic properties of this constraint satisfaction system, using the Necker cube example in *PDP:14* (originally from Feldman, 1981). Ex. 3.2 considers how the schema model deals with knowledge that has traditionally been taken as evidence for schemata, using the room example in *PDP:14*. Exs. 3.3 and 3.4 are more complex projects: Ex. 3.3 suggests you try the tic-tac-toe example in *PDP:14* and Ex. 3.4 is even more open ended. In Appendix E we provide answers for the questions in Exs. 3.1 and 3.2.

Ex. 3.1. The Necker Cube

Feldman (1981) has provided a clear example of a constraint satisfaction problem well-suited to a PDP implementation. That is, he has shown how a simple constraint satisfaction model can capture the fact that there are exactly two good interpretations of a Necker cube. In *PDP:14* (pp. 8-17), we describe a variant of the Feldman example relevant to this exercise. In this example we assume that we have a 16-unit network (as illustrated in Figure 1). Each unit in the network represents a hypothesis about the correct interpretation of a vertex of a Necker cube. For example, the unit in the lower left-hand part of the network represents the hypothesis that the lower left-hand vertex of the drawing is a front-lower-left (*fll*) vertex. The upper right-hand unit of the network represents the hypothesis that the upper right-hand vertex of the Necker cube represents a front-upper-right (*fur*) vertex. Note that these two interpretations are inconsistent in that we do not normally see both of those vertices as being in the frontal plane. The Necker cube has eight vertices, each of which has two possible interpretations—one corresponding to each of the two interpretations of the cube. Thus, we have a total of 16 units. Three kinds of constraints are represented in the network. First, since each vertex can have only one interpretation, we have a negative connection between units representing alternative interpretations of the same vertex. Second, since the same interpretation cannot be given to more than one vertex, units representing the same interpretation are mutually inhibiting. Finally, units that represent locally consistent interpretations should be mutually exciting. Thus, there are positive connections between a unit and its three consistent neighbors. The system will achieve maximal goodness when all units representing one consistent interpretation of the Necker cube are turned on

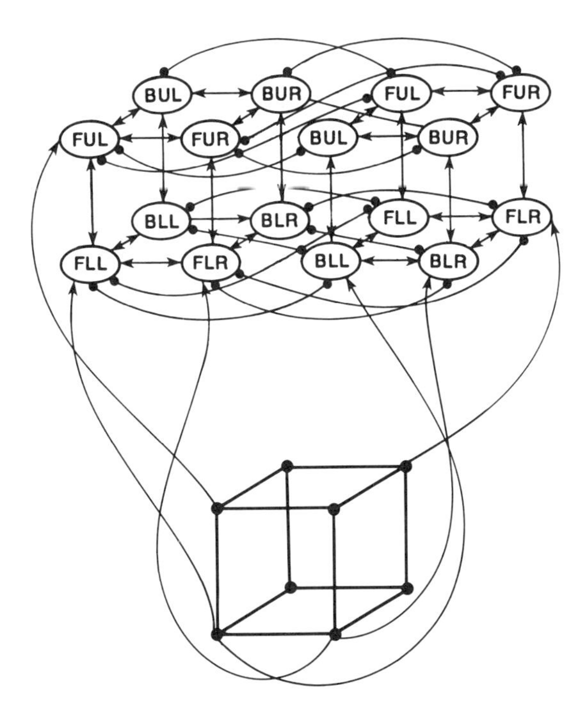

FIGURE 1. A simple network representing some of the constraints involved in perceiving the Necker cube. (From *PDP:14*, p. 10.)

and those representing the other interpretation are turned off. In the diagram, the two subsets of units are segregated so that we expect that either the eight units on the left will come on with the others turned off or the eight units on the right will come on.

Using the *cube.tem* and *cube.str* files, explore the Necker cube example described in *PDP:14*. Run the program several times and look at the obtained interpretations. Record the distribution of interpretations.

Start up the **cs** program with the *cube* template and start-up files:

cs cube.tem cube.str

At this point the screen should look like the one shown in Figure 2. The display depicts the two interpretations of the cube and shows the activation values of the units, the current cycle number, the current update number, the name of the most recently updated unit (there is none yet so this is blank), the current value of goodness, and the current temperature. (The temperature is irrelevant for this exercise, but will become important later.) The activation values of all 16 units are shown, initialized to 0, at the corners of the two cubes drawn on the screen. The units in the cube on the left, cube A, are the ones consistent with the interpretation that the cube is facing down and to the left. Those in the cube on the right, cube B, are the ones consistent with the interpretation of the cube as facing up and to the right. The dashed lines do not correspond to the connections among the units, but simply indicate the interpretations of the units. The connections are those shown in the Necker cube network in Figure 1. The vertices are labeled, and the labels on the screen correspond to those in Figure 1. All units have names. Their names are given by a capital letter

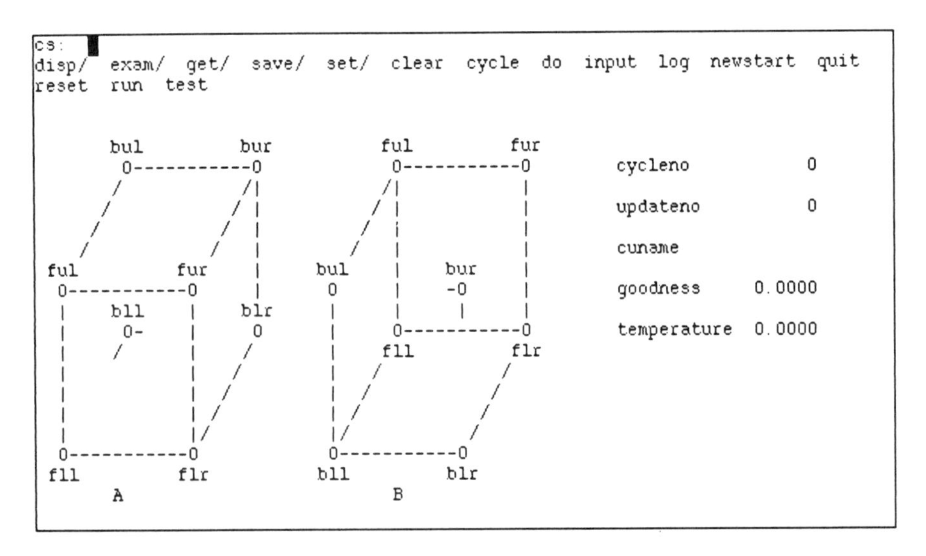

FIGURE 2. The initial image of the screen for the cube example.

indicating which interpretation is involved (A or B), followed by the label appropriate to the associated vertex. Thus, the unit displayed at the lower left vertex of cube A is named *Afll*, the one directly above it is named *Aful* (for the front-upper-left vertex of cube A), and so on. You can use these names to examine the activation values and connections among the units. Thus, for example, it is possible to examine the connection between the unit *Aflr* (the unit representing the hypothesis that the lower right-hand vertex of the Necker cube is in the frontal plane—interpretation A) and the unit *Bblr* (the unit representing the hypothesis that the lower right-hand vertex of the Necker cube is in the back plane—interpretation B) by giving the names *Aflr* and *Bblr* when examining weights. The weights between these two units should be −1.5. (This is reasonable since these two units represent alternative interpretations of the same vertex and so should be inhibitory.)

We are now ready to begin exploring the cube example. The biases and connections among the units have already been read into the program (they were specified in the *cube.net* file, read in by the *get/ network* command in the *cube.str* file). In this example, all units have positive biases, therefore there is no need to specify inputs. Simply type *cycle*. After the command is typed, the display will flash, and various numbers representing the activation values of the corresponding units will replace the 0s at the corners of the cubes. Only single digits are displayed. The numbers are the tenths digit of the activation levels, so that a 4 in the display indicates that the corresponding unit's activation is between 0.4 and 0.5. When the activation values reach 1.0 (their maximum value), an asterisk is plotted. After the display stops flashing you should see that the variables on the right have attained some new values, and you should have a display roughly like that in Figure 3. The variable *cycle* should be 20, indicating that the program has completed 20 cycles. The variable *update* should be at 16, indicating that we have completed the 16th update of the cycle. The *uname* will indicate the last unit updated. The *goodness* may have a value of 6.4. If it does, the network has reached a *global* maximum and has found one of the two "standard" interpretations of the cube. In this case you should find that the activation values of those units in one subnetwork have all reached their maximum value (indicated by asterisks) and those in the other subnetwork are all at 0. If the goodness value is less than 6.4, then the system has found a *local maximum* and there will be nonzero activation values in both subnetworks. You can run the cube example again by issuing the *newstart* command and then entering *cycle*. Do this, say, 20 times to get a feeling for the distribution of final states reached.

Q.3.1.1. How many times was each of the two valid interpretations found? How many times did the system settle into a local maximum? What were the local maxima the system found? Do they correspond to reasonable interpretations of the cube?

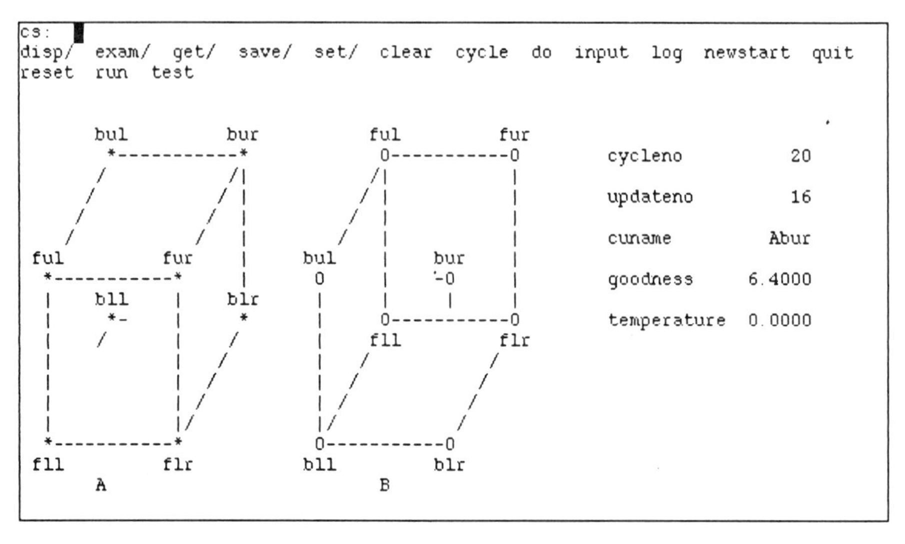

FIGURE 3. The state of the system after 20 cycles.

Now that you have a feeling for the range of final states that the system can reach, try to see if you can understand the course of processing leading up to the final state.

Q.3.1.2. What causes the system to reach one interpretation or the other? How early in the processing cycle does the eventual interpretation become clear? What happens when the system reaches a local maximum? Is there a characteristic of the early stages of processing that leads the system to move toward a local maximum?

Hints. The movement on the screen can be rapid, and it may be difficult to see exactly what is happening. It is sometimes useful to set *single* to 1 and to set *stepsize* to *update*. Under these conditions, the program refreshes the display and pauses after each update. Note also that if you wish to study the evolution of the system toward a particular end state, you can issue the *newstart* command repeatedly, followed by *cycle*, with *single* set to 0, until the system settles to the desired end state, and then use *reset* to repeat the identical run, perhaps first setting *single* to 1 and *stepsize* to *update*.

Q.3.1.3. There is a parameter in the schema model, *istr,* that multiplies the weights and biases and that, in effect, determines the rate of activation flow within the model. The probability of finding a local maximum depends on the value of this parameter. How does the relative frequency of local maxima vary as this parameter is varied? Try several values from 0.08 to 2.0. Explain the results you obtain.

Hints. You will probably find that at low values of *istr* you will want to increase *ncycles*. Alternatively, you can just issue the *cycle* command a second time if the network doesn't settle in the first 20 cycles. You may also want to set *single* back to 0 and *stepsize* back to *cycle*. Do not be disturbed by the fact that the values of *goodness* are different here than in the previous runs. Since *istr* multiplies the weights, it also multiplies the goodness so that *goodness* is proportional to *istr*.

Reset *istr* to its initial value of 0.4 before proceeding to the next question.

Q.3.1.4. It is possible to use external inputs to bias the network in favor of one of the two interpretations. Study the effects of adding an input of 1.0 to the units in one of the subnetworks, using the *input* command. Look at how the distribution of interpretations changes as a result of the number of units receiving external input in a particular subnetwork.

Ex. 3.2. Schemata for Rooms

Interestingly, a simple constraint satisfaction network of the type we have just been describing leads to an interesting interpretation of what a *schema* may be like and how schemata may be implemented in PDP networks. This idea was described in some detail in *PDP:14*. Here we offer a brief summary. The basic idea is that our knowledge is in the form of a constraint satisfaction network. Conventionally (cf. Rumelhart & Ortony, 1977), a schema is a higher-level conceptual structure for representing the complex relationships implicit in our knowledge base. Schemata are data structures for representing generic concepts stored in memory. They are like models of the outside world. Information is processed by first finding the schema that best fits the current situation and then using that model to *fill in* aspects of the situation not specified by the current input. In general, a consistent configuration of such models (or schemata) is discovered that constitutes an interpretation of the situation in question. Within the PDP framework there is no explicit unit or other representational entity corresponding to a schema. Rather, schemata are implicit in our knowledge and arise, while processing information, from the interactions of a large set of constraints. Thus, the units of such a constraint network correspond to hypotheses that certain semantic features are appropriate descriptions of the situation in question. Some of these features are available in the input and form the starting place of the interpretation process. Others are unspecified and must be *filled in* during the process of interpretation. The final state achieved by the network corresponds to the interpretation. Thus, interpretations correspond to local maxima defined over the "goodness" landscape. The constraint satisfaction network captures the two most basic characteristics of schemata: It has a goodness of fit measure and finds the interpretation that is maximally consistent with both the external constraints and the internal knowledge. It automatically completes the pattern and thereby fills in unspecified variables in the situation with their default values.

The primary example in *PDP:14* was an illustration of how a network constructed according to these simple principles could be a constraint network that behaved as if it contained schemata for five different kinds of rooms—for living rooms, kitchens, bedrooms, offices, and bathrooms. The units in this case stood for the hypotheses that a particular room in question contained a typewriter, coffee cup, sofa, bed, and so on. In all, 40 such features were considered. These features are shown in Table 1. The connection strengths themselves were derived from the co-occurrences of the particular characteristics generated from a simple experiment in which a subject imagined rooms of different types and then judged for all 40 features whether or not they were true of the rooms she was imagining. If two features generally occurred together, the connection between them was strong; if two features occurred separately, the connections between them were negative. The details are described in *PDP:14*.

To begin exploring the issues, use the *room.tem* and *room.str* files to replicate the basic five prototypes illustrated on pages 26 and 27 of *PDP:14*. As in the text, use *oven, desk, bathtub, sofa,* and *bed* as the seeds for the prototypes.

We can find the prototype kitchen, for example, by clamping the *oven* unit and letting the system settle to a solution, filling in the unspecified features.[1] You can use the *input* command to set the value of *oven* to 1 and thereby clamp it on. Enter *cycle*; this will cause 50 cycles to be run, since

TABLE 1

THE 40 ROOM DESCRIPTORS

ceiling	walls	door	windows	very-large
large	medium	small	very-small	desk
telephone	bed	typewriter	bookshelf	carpet
books	desk-chair	clock	picture	floor-lamp
sofa	easy-chair	coffee-cup	ashtray	fireplace
drapes	stove	coffeepot	refrigerator	toaster
cupboard	sink	dresser	television	bathtub
toilet	scale	oven	computer	clothes-hanger

(From *PDP:14*, p. 22.)

[1] In *PDP:14*, we clamped on the *ceiling* unit as well as one other unit. Here, we have set the biases on the *ceiling* and *wall* units that essentially force these units to come on, saving you the trouble of clamping one of them in the input command.

ncycles is set to 50. At this point, the system should almost always reach the state shown in Figure 4, with a goodness of about 21.2. The screen display gives the names of each of the units in the network. The external input to each unit is indicated to the left of the unit's name; its activation is indicated to the right. External inputs of +1.0 are indicated by two asterisks; inputs of −1.0 are indicated by two asterisks in reverse video. (Note that in this example, unlike the Necker cube example, the *clamp* variable is set to 1. Thus, units whose inputs are set to a positive number will be forced to stay on.) Activations and other values of external input are indicated in hundredths, so, for example, the activation of the *drapes* unit is 0.99 and the activation of the *oven* unit is 1.0.

Q.3.2.1. Does the system always settle to the same pattern for each prototype? How do the final values of goodness differ for the different prototypes? Why do they differ? Do other initial inputs lead to other patterns? Are there other prototypes in the network that can be accessed from clamping a single unit? Try a couple of runs replicating the results shown on page 34 of *PDP:14* when *bed* and *sofa* have been clamped on. What happens when you clamp other pairs of units, like *sofa* and *bathtub*?

Hints. It is better to use the *newstart* command rather than *reset*. This ensures that successive runs are independent of each other. Note that *newstart* should be called *after* changing the external input with the *input* command. To answer these questions, it will be useful to refer to the weights used in the room example. These are shown in Figure 5.

```
cs:
disp/  exam/  get/  save/  set/  clear  cycle  do  input  log  newstart  quit
reset  run  test

 0 ceiling 100   0 very-sm   0   0 desk-ch   0   0 fire-pl   0   0 dresser   0
 0 walls   100   0 desk      0   0 clock   100   0 drapes   99   0 televis   0
 0 door      0   0 telepho  98   0 picture   0   0 stove   100   0 bathtub   0
 0 window  100   0 bed       0   0 floor-l   0   0 sink    100   0 toilet    0
 0 very-la   0   0 typewri   0   0 sofa      0   0 refrige 100   0 scale     0
 0 large     0   0 book-sh   0   0 easy-ch   0   0 toaster 100   0 coat-ha   0
 0 medium    0   0 carpet    0   0 coffee-  99   0 cupboar 100   0 compute   0
 0 small    99   0 books     0   0 ash-tra   0   0 coffeep 100  ** oven    100

  cycleno  50     goodness 21.2014     temperature  0.0000
```

FIGURE 4. The kitchen prototype, as seen in the **cs** program's display.

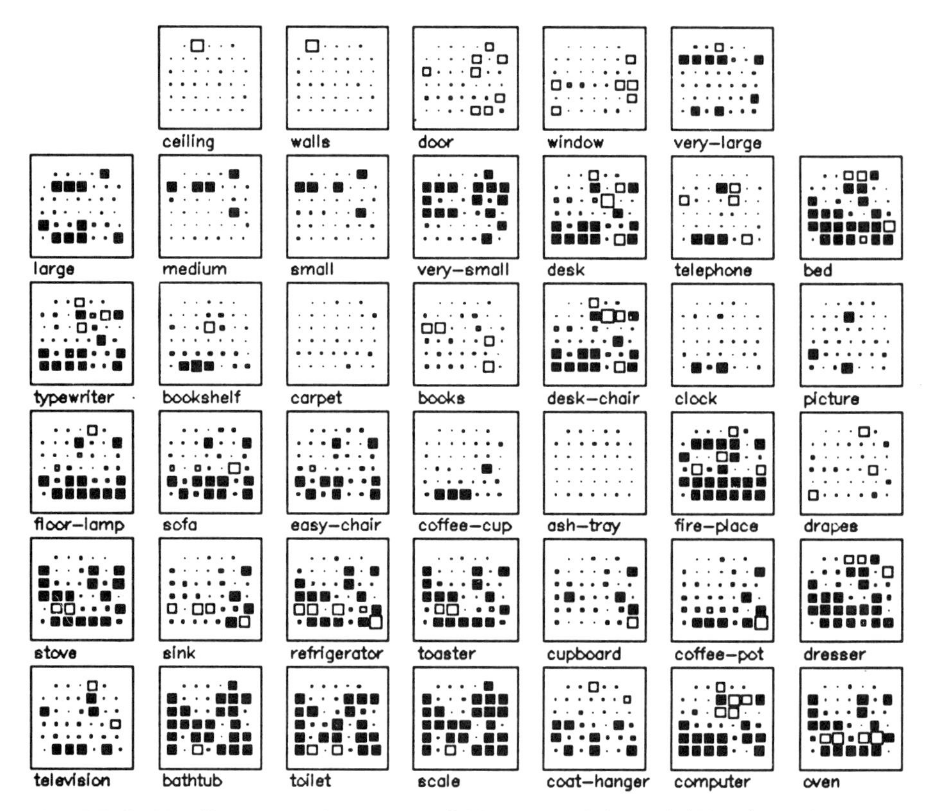

FIGURE 5. The figure uses the method of Hinton and Sejnowski (*PDP:7*) to display the weights. Each unit is represented by a square. The name below the square names the descriptor represented by each square. Within each unit, the small black and white squares represent the weights from that unit to each of the other units in the system. The relative position of the small squares within each unit indicates the unit with which that unit is connected. (From *PDP:14*, p. 24.)

On page 35 of *PDP:14* we illustrate a case in which the goodness is greater for *office* when the units *window* and *drapes* are both on or are both off than when only one of them is on. Using the program, find the numeric values of the goodnesses for the four combinations of *office* with or without windows and with or without drapes. The point of this example is that, in the case of *office*, window and drapes form a kind of a unit. An office is better with both or neither than with one alone. The window-drapes cluster interacts differently with different room types.

Finding the goodness values is a bit tricky. Essentially, you have to clamp all of the units to their prototype values and cycle a few times. Setting all 40 values using *input* can be tedious. In this case, the simplest way to proceed is to get a set of patterns from a file and then, using *test*, select one of the patterns and let the system cycle a few times to compute the goodness. For this example, there is a file called *room.pat* that contains the patterns for the prototype of each of the five rooms. This can be accessed through the *get/ patterns* command. The following interchange will access the patterns.

> **cs:** *get patterns*
> **filename for patterns:** *room.pat*

The patterns for the prototypes are named *office*, *living*, *bedroom*, *bathroom*, and *kitchen*. You can examine these patterns with the command *display/ env*. It is a good idea to clear the screen first. The *display/ env* command produces a display that lists the names of the defined patterns down the left side of the screen and the names of the units (printed vertically) across the top. The patterns are rows of 1s and −1s. For example, in the office pattern, the units *ceiling*, *wall*, *door*, *large*, *desk*, *telephone*, *typewriter*, and several others are on.

You can test the system using the *office* pattern with the *test* command. It is a good idea to set *ncycles* to a small value since the final goodness value for the entire clamped network will be reached almost immediately. We are now in a position to try the window-drapes example. First test the *office* pattern. Note that both the *window* and the *drapes* units are off and that the goodness is 23.78. Now, turn *drapes* on without disturbing the rest of the inputs. This can be done with the *input* command, although you must be careful not to reset all of the input values. Now, reset the activation levels of the units and cycle.[2] You should find a goodness level of 23.11. Now clamp the *window* unit on, so that both the *window* and *drapes* units are clamped on, reset and cycle again. In this case with *both* units on, we have a goodness of 23.62. Finally, turn the *drapes* unit off (i.e., give it a negative input value), so that the *window* unit is on and *drapes* is off, and enter *reset* and *cycle* again. Here you should find a goodness of 23.35. You should now be able to carry out this procedure with the rest of the prototypes.

Q.3.2.2. First repeat the window experiment with the bathroom prototype. In what way are the results different? Try it with bedroom, kitchen, and living room prototypes as well. Note the different patterns of results in the different contexts. What sense can you make of these different patterns?

Ex. 3.3. Jets and Sharks

An additional example that can be studied within the context of the **cs** program is the Jets and Sharks example from Chapter 2. The appropriate

---

2 Note that the *reset* and *newstart* commands are both fine to use in this case because all units are clamped and so the order of updating does not matter.

files are all set up; you can explore the network's performance on many of the same inputs we used in Chapter 2. To run, simply start up the **cs** program with the *jets.tem* and *jets.str* files, specify external inputs as appropriate, and then use the *cycle* command. See if you can understand and explain the differences between the way the two models behave on these two examples.

In experimenting with this example, it is interesting to separately explore the effects of varying the inhibition on the input units and the hidden units. This must be done by changing the values assigned to the constraint letters *h* (for hidden) and *v* (for visible) in the *jets.net* file. As supplied, *v* is set to −2.0 and *h* to −1.0. Thus, inhibition is relatively strong among the visible units, but not so strong among hidden (instance) units. You can change these values by editing this file if you wish to explore this matter further.

Ex. 3.4. Tic-Tac-Toe

As a final suggestion for an experiment to try with the schema model, we propose the following somewhat more challenging exercise. Using the basic design illustrated on page 49 of *PDP:14*, build a version of the schema model that is able to take a tic-tac-toe board as an input and settle into a state in which it produces the appropriate move. This will involve building an appropriate *tic.tem* file, an appropriate *tic.str* file, and appropriate *tic.net* and *tic.wts* files.

LOCAL MAXIMA AND THE PHYSICS ANALOGY

In this section we provide a brief description of Hinton and Sejnowski's Boltzmann machine and of Smolensky's harmony theory as they are described in *PDP:7* and *PDP:6*, respectively. These systems were developed from an analogy with statistical physics and it is useful to put them in this context. We thus begin with a description of the physical analogy and then show how this analogy solves some of the problems of the schema model described earlier. Then we turn to a description of the Boltzmann machine, show how it is implemented, show how the **cs** program can be used in *boltzmann* mode to solve constraint satisfaction problems, and finally discuss harmony theory and how it can be implemented using the **cs** program.

The primary advantage of these systems over the deterministic constraint satisfaction system used in the schema model is their ability to overcome the problem of local maxima in the goodness function. It will be useful to begin with an example of a local maximum and try to understand in some detail why it occurs and what can be done about it. Figure 6 illustrates a typical example of a local maximum with the Necker cube. Here we see that the system has settled to a state in which the lower four vertices were organized according to interpretation A and the upper four vertices were organized according to interpretation B. Local maxima are always blends of parts of the two global maxima. We never see a final state in which the points are scattered randomly across the two interpretations. All of the local maxima are cases in which one small cluster of adjacent vertices are organized in one way and the rest are organized in another. This is because the constraints are local. That is, a given vertex supports and receives support from its neighbors. The units in the cluster mutually support one another. Moreover, the two clusters are always arranged so that none of the inhibitory connections are active. Note in this case, *Bfur* is on and the two units it inhibits, *Afur* and *Abur*, are both off. Similarly, *Bbur* is on and *Abur* and *Afur* are both off. Clearly the system has found little coalitions of units that hang together and conflict minimally with the other coalitions.

In Q.3.1.2 of Ex. 3.1, you had the opportunity to explore the process of settling into one of these local maxima. What happens is this. First a unit in one subnetwork comes on. Then a unit in the other subnetwork, which does not interact directly with the first, is updated, and, since it has a positive bias and at that time no conflicting inputs, it also comes on. Now the next unit to come on may be a unit that supports either of the two units already on or possibly another unit that doesn't interact directly with either of the other two units. As more units come on, they will fit into one or another of these two emerging coalitions. Units that are directly inconsistent with active units will not come on or will come on weakly and then probably be turned off again. In short, local maxima occur when units that

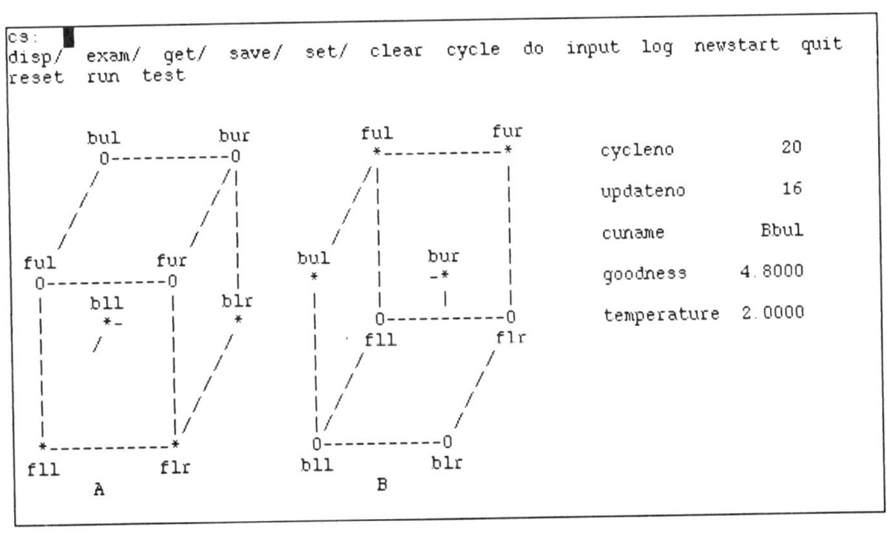

FIGURE 6. A local minimum with the Necker cube.

don't interact directly set up coalitions in both of the subnetworks; by the time interaction does occur, it is too late, and the coalitions are set.

Interestingly, the coalitions that get set up in the Necker cube are analogous to the bonding of atoms in a crystalline structure. In a crystal the atoms interact in much they same way as the vertices of our cube. If a particular atom is oriented in a particular way, it will tend to influence the orientation of nearby atoms so that they fit together optimally. This happens over the entire crystal so that some atoms in one part of the crystal can form a structure in one orientation while atoms in another part of the crystal can form a structure in another orientation. The points where these opposing orientations meet constitute flaws in the crystal.

It turns out that there is a strong mathematical similarity between our network models and these kinds of processes in physics. Indeed, the work of Hopfield (1982, 1984) on so-called Hopfield nets, of Hinton and Sejnowski (1983, *PDP:7*) on the Boltzmann machine, and of Smolensky (1983, *PDP:6*) on harmony theory were strongly inspired by just these kinds of processes. In physics, the analogs of the goodness maxima of the above discussion are *energy* minima. There is a tendency for all physical systems to evolve from highly energetic states to states of minimal energy.

In 1982, John Hopfield, a physicist, observed that symmetric networks using deterministic update rules behave in such a way as to minimize an overall measure he called *energy* defined over the whole network. Hopfield's energy measure was essentially the negation of our goodness measure. We use the term *goodness* because we think of our system as a system for maximizing the goodness of fit of the system to a set of constraints. Hopfield, however, thought in terms of energy, because his networks behaved very much as thermodynamical systems, which seek minimum energy states. In physics the stable minimum energy states are called *attractor states.* This analogy of networks falling into energy minima just as physical systems do has provided an important conceptual tool for analyzing parallel distributed processing mechanisms.

Hopfield's original networks had a problem with local "energy minima" that was much worse than in the schema model described earlier. His units were binary. (Hopfield, 1984, has since gone to a version in which units take on a continuum of values to help deal with the problem of local minima in his model. The schema model is similar to Hopfield's 1984 model.) For binary units, if the net input to a unit is positive, the unit takes on its maximum value; if it is negative, the unit takes on its minimum value (otherwise, it doesn't change value). Binary units are more prone to local minima because the units do not get an opportunity to communicate with one another before committing to one value or the other. In Q.3.1.3 of Ex. 3.1, we gave you the opportunity to run a version close to the Hopfield model by setting *instr* to 2.0 in the Necker cube example. In this case the units are always at either their maximum or minimum values. Under these conditions, the system reaches local goodness maxima (energy minima in Hopfield's terminology) much more frequently.

Once the problem has been cast as an energy minimization problem and the analogy with crystals has been noted, the solution to the problem of local goodness maxima can be solved in essentially the same way that flaws are dealt with in crystal formation. One standard method involves *annealing*. Annealing is a process whereby a material is heated and then cooled very slowly. The idea is that as the material is heated, the bonds among the atoms weaken and the atoms are free to reorient relatively freely. They are in a state of high energy. As the material is cooled, the bonds begin to strengthen, and as the cooling continues, the bonds eventually become sufficiently strong that the material freezes. If we want to minimize the occurrence of flaws in the material, we must cool slowly enough so that the effects of one particular coalition of atoms has time to propagate from neighbor to neighbor throughout the whole material before the material freezes. The cooling must be especially slow as the freezing temperature is approached. During this period the bonds are quite strong so that the clusters will hold together, but they are not so strong that atoms in one cluster might not change state so as to line up with those in an adjacent cluster—even if it means moving into a momentarily more energetic state. In this way annealing can move a material toward a global energy minimum.

The solution then is to add an annealing-like process to our network models and have them employ a kind of *simulated annealing*. The basic idea is to add a global parameter analogous to temperature in physical systems and therefore called *temperature*. This parameter should act in such a way as to decrease the strength of connections at the start and then change so as to strengthen them as the network is settling. Moreover, the system should exhibit some random behavior so that instead of always moving uphill in goodness space, when the temperature is high it will sometimes move downhill. This will allow the system to "step down from" goodness peaks that are not very high and explore other parts of the goodness space to find the global peak. This is just what Hinton and Sejnowski have proposed in the Boltzmann machine, what Geman and Geman (1984) have proposed in the *Gibbs sampler*, and what Smolensky has proposed in harmony theory.

The essential update rule employed in all of these models is probabilistic and is given by what we call the *logistic* function:

$$probability\,(a_i(t) = 1) = \frac{1}{1 + e^{-net_i/T}} \tag{8}$$

where T is the temperature. This differs from the basic schema model in three important ways. First, like Hopfield's original model, the units are binary. They can take on only their maximum and minimum values. Second, they are *stochastic*. That is, the update rule specifies only a *probability* that the units will take on one or the other of their values. This means that the system need not necessarily go uphill in goodness—it can move downhill as well. Third, the behavior of the systems depends on a global parameter, *temperature*, which can start out high and be reduced during the settling phase. These characteristics allow these systems to implement a simulated annealing process. One final point: It is not accidental that these three models all choose exactly the same update rule. This rule is drawn directly from physics and there are important mathematical results that, in effect, guarantee that the system will end up in a global maximum if the system is annealed slowly enough. Having made the analogy with physics, we can also make use of the results of physics to describe the behavior of our networks.

Figure 7 shows the probability values as a function of net input and the temperature. Several observations should be made. First, if the net input is 0, the unit takes on its maximum and minimum values with equal probability. Second, if the net input is large enough, the unit will always take on its maximum value no matter what value the temperature is; and if the net input in sufficiently negative, the unit will take on its minimum value no matter what the temperature. Third, as the temperature approaches 0, the function becomes deterministic and takes on its maximum value if the net input is positive and its minimum value if the net input is negative. This "zero temperature" case is identical to the Hopfield binary unit model. Thus, we see that there are two dimensions on which these models are varying—whether the units are binary or continuous values and whether the models are stochastic (probabilistic) or deterministic. The Hopfield (1982) model is deterministic and binary. The Boltzmann machine and harmony theory are stochastic and binary. The Hopfield (1984) model and the schema model are deterministic and continuous. All five models converge to binary, deterministic models as the temperature goes to 0 in the

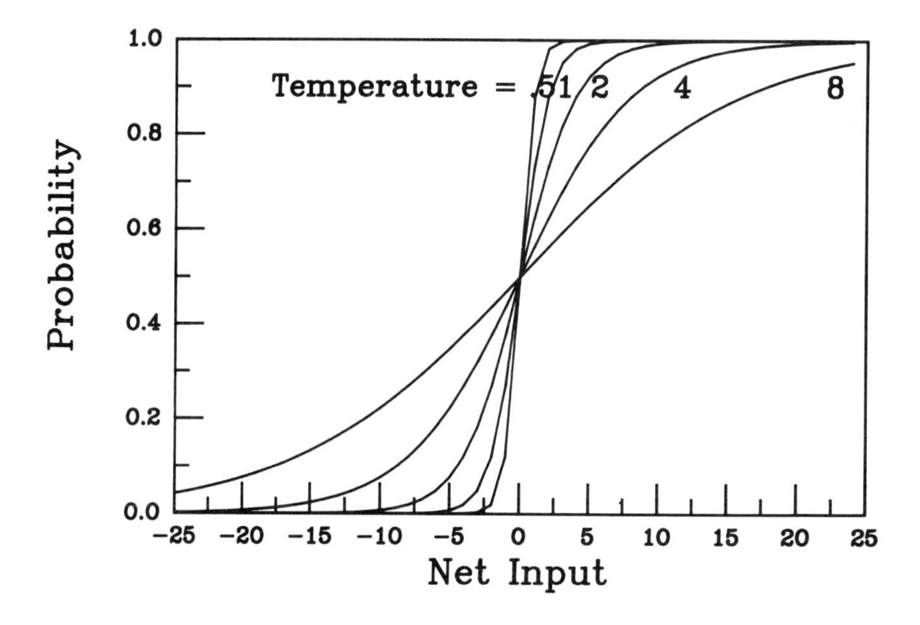

FIGURE 7. Activation probabilities according to the logistic function. (From *PDP:2*, p. 69.)

stochastic models and as the size of the step taken per update (controlled by *istr*) is increased in the schema model.

THE BOLTZMANN MACHINE

In *PDP:7* Hinton and Sejnowski described the Boltzmann machine as a constraint satisfaction system. Most of their chapter, however, focuses on the development of a very interesting learning procedure. Here, we focus on the use of the Boltzmann machine as a constraint satisfaction system. We will especially focus on its role in reducing the frequency of local goodness maxima. (Following Hopfield and the physics analogy, Hinton and Sejnowski spoke of energy and energy minimization. We will persist in looking at the negation of energy, which we call goodness. The principles are identical in either case, only the terminology varies.)

Implementation

The Boltzmann machine is conceptually very similar to the schema model. Indeed, the Boltzmann system is accessed as a mode of the **cs** program. As in the schema model, there are two essential subroutines for Boltzmann. These are a *cycle* and an *rupdate* routine. The *cycle* routine is absolutely identical for the two models. The *rupdate* routine differs by only a few lines of code—namely, the code for assigning an activation value to a unit. Since they are nearly the same, we have deleted the comments and reproduced the code for the Boltzmann version of *rupdate* below.

```
rupdate() {

  for (updateno = 0; updateno < nupdates; updateno++) {
    i = randint(0, nunits - 1);
    netinput = 0;
      for(j = 0; j < nunits; j++) {
        netinput += activation[j]*weight[i][j];
      }
      netinput += bias[i];
      netinput *= istrength;
      netinput +- estrength * input[i];

      if(probability(logistic(netinput)) == 1)
        activation[i] = 1;
      else
        activation[i] = 0;
  }
}
```

where *logistic(x)* is a function given by

```
logistic(x) {

  return(1.0 / (1.0 + exp( -1 * x / temperature)));
}
```

and *probability(x)* is a function that takes on value 1 with a probability equal to the value of its argument:

```
probability(x) {

  if (rnd() < x)
    return(1);
  else
    return(0);
}
```

The *rnd()* function simply returns a uniformly distributed random number between 0 and 1.

Running the Program in Boltzmann Mode

Since the Boltzmann machine is so similar to our previous program, it is implemented as mode of program **cs**. It is accessed by setting *mode boltzmann* to 1. All of the commands and variables needed are described in the list that begins on page 56. Of these, the most important specifically for Boltzmann machines is the *get/ annealing* command, used to specify an annealing schedule.

Ex. 3.5. Simulated Annealing in the Cube Example

Compare simulated annealing using the Boltzmann machine to the results you obtained with the schema model for the cube example. To switch to Boltzmann mode, enter *set/ mode boltzmann 1* after starting up the program with *cube.tem* and *cube.str*. You may also want to increase *ncycles* to 60, since this is just beyond the point at which the annealing schedule levels off.

Q.3.5.1. Run 20 runs with the Boltzmann machine, using its default annealing schedule, and see how often the program gets stuck in a local minimum. Compare these results to the case where the continuous updating scheme is used.

Q.3.5.2. Examine the density of local maxima as a function of the annealing schedule. Try several schedules varying in gradualness and starting temperature. Which variable makes more of a difference? Can you explain why?

Hints. It is necessary to run 20 runs or so with each schedule to get reasonable estimates of the probabilities of getting stuck in local maxima. Use relatively high values of the starting temperature and very quick drops if you want to get a feel for what is happening. To study the time course of these simulations, it is useful to set *stepsize* to *update* and to set *single* to 1. Then you can step through and watch the network settle. You will note that things begin to settle down when the temperature begins to get rather low, say, in the range of 0.5 to 0.05.

Other Experiments With Boltzmann Machines

There are many more experiments that can be done with Boltzmann machines. For example, you may explore the effects of stochastic variability in the room example, the Jets and Sharks example, or the tic-tac-toe example. We provide one additional example, called *boltz*. To run it, execute the **cs** program with the *boltz.tem* and *boltz.str* files. The screen display indicates the excitatory connections; you will want to study the *boltz.wts* file to see what the inhibitory connections are. You can also construct examples of your own, perhaps setting up a network that you feel might be challenging for finding global minima reliably.

HARMONY THEORY

Harmony theory, which was developed by Paul Smolensky, is described in *PDP:6*. Although the basic mathematics of harmony theory is rather similar to the Boltzmann machine, the structure and motivation are different. Whereas we can think of the Boltzmann machine as an arbitrarily interconnected set of homogeneous units, harmony theory presupposes two distinct layers of units. As illustrated in Figure 8, a harmony network consists of a lower layer of *representational feature* units and an upper layer of *knowledge atoms.* The feature units take on activation values ± 1, whereas the knowledge atoms take on values 0 and 1. It is useful to think of the feature units as corresponding to the featural description of a situation. In a complete description, each feature is either present (+1) or absent (−1). The knowledge atoms, on the other hand, are best thought of as bits of knowledge about what configurations of features "go together." Knowledge

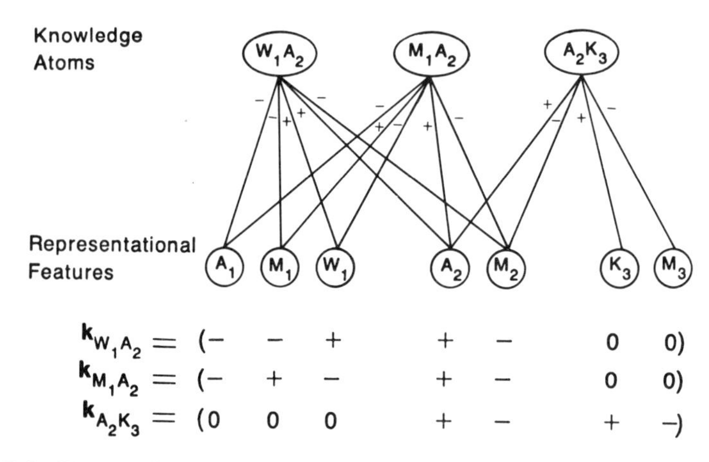

FIGURE 8. The graphical representation of a particular harmony model. In this model, the representational features represent letters in particular positions, and the knowledge atoms represent common letter combinations. (From *PDP:6*, p. 216.)

atoms may be either active or inactive. When a knowledge atom is active, it can be viewed as asserting that the configuration of input features it is looking for is present in the environment. When a knowledge atom is inactive it can be viewed as asserting that the evidence does not warrant such an assertion. All connections in a harmony model are symmetric, and all connections are between features and knowledge atoms. Thus, a given feature may either excite a knowledge atom that is consistent with it, inhibit a knowledge atom that is inconsistent with it, or have no effect on a knowledge atom to which the feature is irrelevant. Similarly, knowledge atoms specify certain configurations of features that are consistent with the knowledge represented by that atom. Thus a knowledge atom may activate those features that are consistent with the atom, inhibit those that are inconsistent, or not connect with those that are irrelevant to the contents of that atom. Neither features nor knowledge atoms are directly connected to one another. All connections in the system are ± 1. However, each knowledge atom has a *strength* designated σ. The strength corresponds to the degree that the knowledge atom in question insists that the features to which it is connected are present in the input.

Harmony theory is so named because, for any configuration of input features, the system finds the configuration of knowledge atoms that is maximally consistent, or harmonious, with the featural constraints. It is useful to see the configuration of active knowledge atoms as an *interpretation* of the input features.

In addition to creating an interpretation of a set of input features, the knowledge atoms themselves can *fill in* missing features in a way that is maximally consistent with those features that are fixed (clamped) and the set of knowledge atoms. This is the so-called *completion* problem. The harmony function itself is very similar to the "goodness" function of the previous sections. It can be written in the following way:

$$harmony = \sum_i \sigma_i a_i h_i.$$

Here, i ranges over the knowledge atoms, and h_i is a measure of the degree to which the current set of feature values is consistent with knowledge atom i. The variable σ_i is a strength or importance value associated with unit i. The variable h_i is given by

$$h_i = \frac{\sum_j r_j k_{ij}}{n_i} - \kappa.$$

Here j ranges over features, r_j is the activation of representational feature j, and n_i is the number of nonzero connections to atom i. The variable k_{ij} is given by

$$k_{ij} = \begin{cases} 1 & \text{if positive connection} \\ -1 & \text{if negative connection} \\ 0 & \text{if no connection.} \end{cases}$$

In other words, the total harmony is given by the sum of contributions of each of the knowledge atoms. If a knowledge atom is not activated ($a_i = 0$), there is no contribution. If it is active ($a_i = 1$), then it contributes an amount that is proportional to the product of its importance, σ_i, and a term representing the consistency of that atom with the current pattern of activation among the representational features. This consistency term, h_i, is the proportion of relevant features that are consistent minus the proportion that are inconsistent, less a constant κ. Consider first the case in which κ is 0. In this case, turning on atom i will contribute a positive amount to the overall harmony of the system whenever the number of consistent features exceeds the number of inconsistent features. If κ is near 1, then it will contribute to the overall harmony only when all, or nearly all, of its features match the template for the atom. The standard goodness function discussed in the Boltzmann model and the schema model corresponds to the $\kappa = 0$ case.[3] Given the motivation of harmony theory, larger values of κ make more sense.

---

[3] Note that if $\kappa = 0$ we can represent the harmony function as

$$harmony = \sum_{ij} a_i r_j w_{ij}$$

where $w_{ij} = \sigma_i k_{ij} / n_i$. This is simply the standard goodness function discussed above.

Implementation

The harmony model is implemented as a mode of the **cs** program. The main differences between *harmony* mode and *boltzmann* mode are

- Only the weights to each atom from each feature are specified. The weight to a feature from an atom is just the weight from the feature to the atom.

- The *rupdate* routine is modified so that there are two versions of the update function, depending on whether the unit being updated is a knowledge atom or a feature.

- The *goodness* measure (now called *harmony*) is computed as just described, taking the importance variables σ_i and κ into account.

Values for σ_i are set by default to 1.0. Other values may be specified in the *.net* file, as described in Appendix C.

Running the Program in Harmony Mode

Harmony mode is accessed simply by setting the *harmony* mode variable to 1. The *clamp* mode is also set to 1. The parameter *kappa* and the configuration variables *nunits* and *ninputs* must be defined. The variable *ninputs* indicates the number of features. The rest of the units are treated as knowledge atoms. The configuration variables are generally defined in the *.net* file; the weights and sigmas are also specified there. Once a set of templates has been specified and the network initialized, the harmony version of the **cs** program is run just like the Boltzmann version: Some features are clamped on (+1) or off (−1), an annealing schedule is defined, the network is reset to initialize the annealing schedule, and then the *cycle* command is entered to initiate processing.

Ex. 3.6. Electricity Problem Solving

Consider the electricity problem described in *PDP:6* (p. 240) and illustrated in Figure 9. This problem, first developed by Riley and Smolensky (1984), illustrates how harmony theory can be employed to solve certain "higher level" problems. In this case, the problem is to determine how different variables in an electrical circuit change when other variables are altered. For example, what happens to the total resistance in the circuit

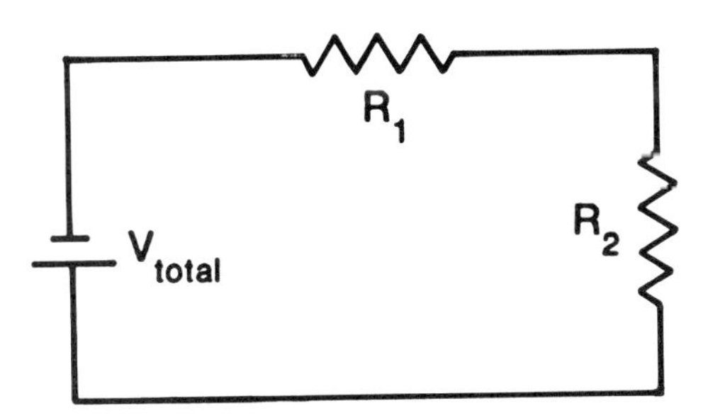

FIGURE 9. If the resistance of R_2 is increased (assuming that V_{total} and R_1 remain the same), what happens to the current and voltage drops across the two resistors? (From *PDP:6*, p. 240.)

shown when the resistance in one of the resistors is increased? Assuming total voltage stays constant, what happens to the voltage drop across each resistor, and what happens to the current? The first step is to develop a set of representational features. In this case, we must represent seven quantities: the total current, I; the resistances, R_1 and R_2; the total resistance, R_{total}; the voltage drops across the two resistors, V_1 and V_2; and the total voltage, V_{total}. For each of these quantities we must represent whether it goes up, goes down, or stays the same. This is done by assigning two units to each quantity: one to indicate whether or not a change occurs in that variable (+1 indicating change and −1 indicating no change) and one to indicate the direction of change. Use +1 to indicate an increase and −1 to indicate a decrease. (If no change occurred, the value of the second feature is irrelevant.) Figure 10 shows the screen layout for the electricity problem. There are columns for each of the seven variables. Below each column is a

```
cs:
disp/  exam/  get/  save/  set/  clear  cycle  do  input  log  newstart  quit
reset  run  test


                  I  R1 R2 RT V1 V2 VT                cycleno         0
       Inputs    00 10 11 00 00 00 10                temp       1.0000
                 cu cu cu cu cu cu cu                harmony    0.0000
       Features  00 00 00 00 00 00 00

                knowledge atom activations

                u u u u u s s s d d d d d
                u s d d d u s d u u u s d
                u u u s d u s d u s d d d

 V1 + V2 = VT   0 0 0 0 0 0 0 0 0 0 0 0 0
 R1 + R2 = RT   0 0 0 0 0 0 0 0 0 0 0 0 0
  I * R1 = V1   0 0 0 0 0 0 0 0 0 0 0 0 0
  I * R2 = V2   0 0 0 0 0 0 0 0 0 0 0 0 0
  I * RT = VT   0 0 0 0 0 0 0 0 0 0 0 0 0
```

FIGURE 10. Screen layout for the electricity problem.

set of pairs of features, one indicating whether or not that quantity changed (indicated by a *c*) and one indicating whether the quantity went up or down (indicated by a *u*). Note that the row labeled *Inputs* is a representation of the problem. The 0s indicate unclamped inputs—inputs to be filled in through processing. The ± 1 values indicate the clamped inputs, which constitute the problem specification. In this case, we have R_2 increasing (both the *change* feature and the *up* feature are clamped $+1$), and we have V_{total} and R_1 unchanged (the *change* feature is clamped to -1). All other features are left free.

The next problem in specifying a harmony network is to encode the knowledge constraints. In this case, the knowledge is of the facts of electrical circuits. We want to represent knowledge about electricity *qualitatively*. We can do this by taking the laws of electricity (Ohm's law and Kirchoff's law), determining the legitimate relationships among the variables involved, and building knowledge atoms for each such relationship. An example should clarify this. Consider first the law that the total voltage drop is the sum of the voltage drops over each resistor, $V_1 + V_2 = V_{total}$. This equation allows for 13 qualitative relationships among the variables. V_1 could increase and V_2 could increase, in which case V_{total} must increase; V_1 could increase and V_2 could stay the same, in which case V_{total} must increase; V_1 could increase and V_2 decrease, in which case V_{total} could increase, stay the same, or decrease; and so on. There are five such equations and 13 qualitative relationships for each equation. This leads to 65 knowledge atoms encoding these relationships. The relationships and knowledge atoms are shown in Figure 10. All of these relationships must be encoded in the network by specifying a positive, negative, or zero weight from each input feature to each knowledge atom. Here is the portion of the network specification for the 13 laws related to voltage:

```
........pppppp
........ppm.pp
........pppmpp
........pppmm.
........pppmpm
........m.pppp
........m.m.m.
........m.pmpm
........pmpppp
........pmppm.
........pmpppm
........pmm.pm
........pmpmpm
```

The *p* represents $+1$ and the *m* represents -1. Since the weights are symmetric in harmony theory, we only require that the connections from the input feature units to the knowledge atom units be specified. Since this problem involves the relationships between voltage, V, current, I, and resistance, R, we have called this the VIR problem and have named the relevant files *vir.tem*, *vir.net*, *vir.str*, and so on.

You should now be ready to run the VIR problem. Note that the inputs specifying our example problem are set up, and an annealing schedule is defined in the *.str* file. All you need to do to run this exercise is start up the **cs** program with the *vir.tem* and *vir.str* files, then issue the *cycle* command.

Q.3.6.1. Watch the system solve the problem and note the final equations selected for the solution. Run the problem several times looking for local maxima.

Hints. This network settles very slowly; it may be necessary to run 300 or sometimes 400 cycles before it is safe to assume things have stopped changing; even then you will find that some of the knowledge atoms flicker on and off. It may be useful to set *stepsize* to *ncycles*, set *ncycles* to 100, and issue the *cycle* command three or four times in each run since much of the total elapsed time can be taken up in screen updates. At the end you may want to run a few more cycles with *single* set to 1 and *stepsize* set to *cycle* to see which units are flickering on and off.

The next two questions are somewhat time consuming and require you to follow the course of processing in the network carefully over time. You will probably have to do several runs in each case to get reliable results. It may be useful to make use of the *do* command facility and log files so that you can run several runs automatically and save the results; Appendix D describes utility programs that may be useful in analyzing the output.

Q.3.6.2. Smolensky reported that the system seemed to come to various conclusions sequentially. See how well you are able to replicate his findings.

Q.3.6.3. Select another problem in the VIR domain (i.e., input a different set of initial constraints) and look at the sequential character of the problem solving in this case. Does it seem reasonable that one might solve the problem in that order? Why does the system seem to settle on different aspects of the problem in different orders?

CHAPTER 4

Learning in PDP Models: The Pattern Associator

In previous chapters we have seen how PDP models can be used as content-addressable memories and constraint-satisfaction mechanisms. PDP models are also of interest because of their learning capabilities. They learn, naturally and incrementally, in the course of processing. In this chapter, we will begin to explore learning in PDP models. We will consider two "classical" procedures for learning: the so-called Hebbian, or correlational learning rule, described by Hebb (1949) and before him by William James (1890), and the error-correcting or "delta" learning rule, as studied in slightly different forms by Widrow and Hoff (1960) and by Rosenblatt (1959).

We will also explore the characteristics of one of the most basic network architectures that has been widely used in distributed memory modeling with the Hebb rule and the delta rule. This is the *pattern associator.* The pattern associator has a set of input units connected to a set of output units by a single layer of modifiable connections that are suitable for training with the Hebb rule and the delta rule. Models of this type have been extensively studied by James Anderson (see Anderson, 1983, for a more recent review), Kohonen (1977), and many others; a number of the papers in the Hinton and Anderson (1981) volume describe models of this type. The models of past-tense learning and of case-role assignment in *PDP:18* and *PDP:19* are pattern associators trained with the delta rule. An analysis of the delta rule in pattern associator models is described in *PDP:11.*

As these works point out, one-layer pattern associators have several suggestive properties that have made them attractive as models of learning and memory. They can learn to act as content-addressable memories; they generalize the responses they make to novel inputs that are similar to the inputs that they have been trained on; they learn to extract the prototype of a set of repeated experiences in ways that are very similar to the concept-learning characteristics seen in human cognitive processes; and they degrade gracefully with damage and noise. In this chapter our aim is to help you develop a basic understanding of the characteristics of these simple parallel networks. However, it must be noted that these kinds of networks have limitations. In the next chapter we will examine these limitations and consider learning procedures that allow the same positive characteristics of pattern associators to manifest themselves in networks and overcome one important class of limitations.

We begin this chapter by presenting a basic description of the learning rules and how they work in training connections coming into a single unit. We will then apply them to learning in the pattern associator.

BACKGROUND

The Hebb Rule

In Hebb's own formulation, this learning rule was somewhat vaguely described. He suggested that when two cells fire at the same time, the strength of the connection between them should be increased. There are a variety of mathematical formulations of this principle that may be given. The simplest by far is the following:

$$\Delta w_{ij} = \epsilon a_i a_j. \tag{1}$$

Here we use ϵ to refer to the value of the learning rate parameter. This version has been used extensively in the early work of James Anderson (e.g., Anderson, 1977). If we start from all-zero weights, then expose the network to a sequence of learning events indexed by l, the value of any weight at the end of a series of learning events will be

$$w_{ij} = \epsilon \sum_l a_{il} a_{jl}. \tag{2}$$

In studying this rule, we will assume that activations are distributed around 0 and that the units in the network have activations that can be set in either of two ways: They may be *clamped* to particular values by external inputs or they may be determined by inputs via their connections to other units in the network. In the latter case, we will initially focus on the case where the units are completely linear; that is, on the case in which the activation and the output of the unit are simply set equal to the net input:

$$a_i = \sum_j a_j w_{ij}. \tag{3}$$

In this formulation, with the activations distributed around 0, the w_{ij} assigned by Equation 2 will be proportional to the correlation between the activations of units *i* and *j*; normalizations can be used to preserve this correlational property when units have mean activations that vary from 0.

The correlational character of the Hebbian learning rule is at once the strength of the procedure and its weakness. It is a strength because these correlations can sometimes produce useful associative learning; that is, particular units, when active, will tend to excite other units whose activations have been correlated with them in the past. It can be a weakness, though, since correlations between unit activations often are not sufficient to allow a network to learn even very simple associations between patterns of activation.

First let's examine a positive case: a simple network consisting of two input units and one output unit (Figure 1A). Suppose that we arrange things so that by means of inputs external to this network we are able to impose patterns of activation on these units, and suppose that we use the Hebb rule (Equation 1 above) to train the connections from the two input units to the output unit. Suppose further that we use the four patterns shown in Figure 1B; that is, we present each pattern, forcing the units to

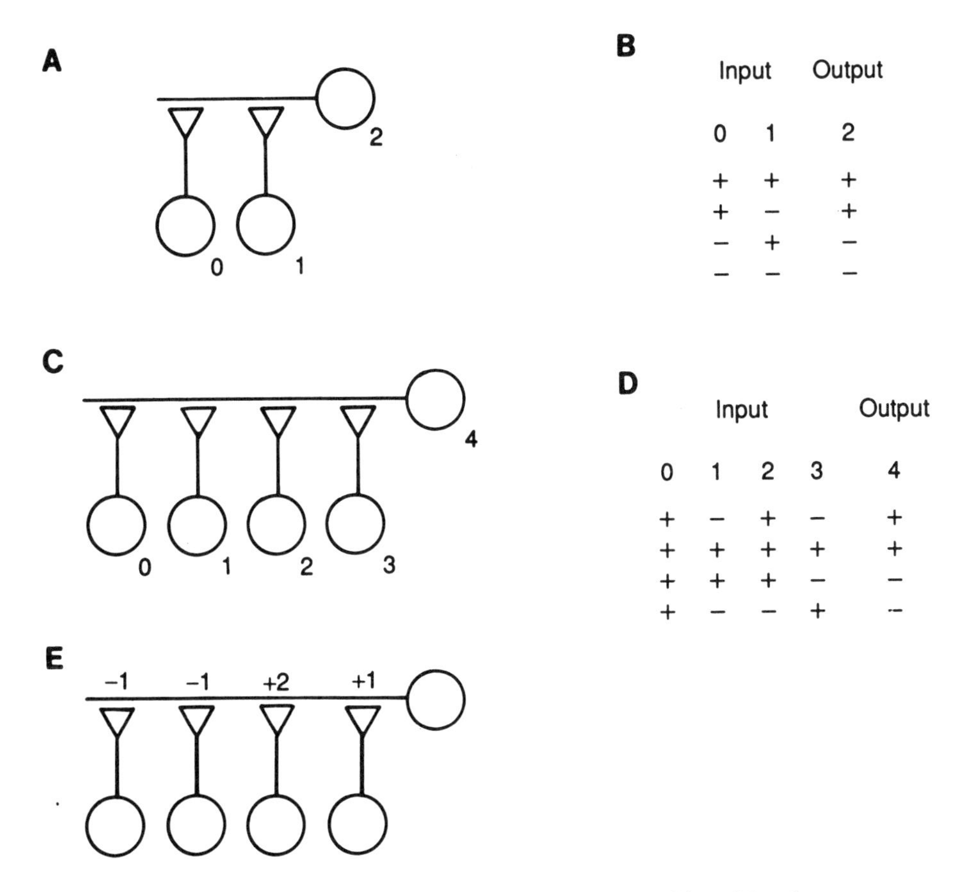

FIGURE 1. Two simple associative networks and the patterns used in training them.

the correct activation, then we adjust the strengths of the connections between the units. According to Equation 1, w_{20} (the weight on the connection to unit 2 from unit 0) will be increased in strength for each pattern by amount ϵ, which in this case we will set to 1.0. On the other hand, w_{21} will be increased by amount ϵ in one of the cases (pattern 0) and reduced by ϵ in the other case, for a net change of 0.

As a result of this training, then, this simple network would have acquired a positive connection weight to unit 2 from unit 0. This connection will now allow unit 0 to make unit 2 take on an activation value correlated with that of unit 0. At the same time, the network would have acquired a null connection from unit 1 to unit 2, capturing the fact that the activation of unit 1 has no predictive relation to the activation of unit 2. In this way, it is possible to use Hebbian learning to learn associations that depend on the correlation between activations of units in a network.

Unfortunately, the correlational learning that is possible with a Hebbian learning rule is a "unitwise" correlation, and sometimes, these unitwise correlations are not sufficient to learn correct associations between whole input patterns and appropriate responses. To see that this is so, suppose we change our network so that there are now four input units and one output unit, as shown in Figure 1C. And suppose we want to train the connections in the network so that the output unit takes on the values given in Figure 1D for each of the four input patterns shown there. In this case, the Hebbian learning procedure will not produce correct results. To see why, we need to examine the values of the weights (equivalently, the pairwise correlations of the activations of each sending unit with the receiving unit). What we see is that three of the connections end up with 0 weights because the activation of the corresponding input unit is uncorrelated with the activation of the output unit. Only one of the input units, unit 2, has a positive correlation with unit 4 over this set of patterns. This means that the output unit will make the same response to the first three patterns since in all three of these cases the third unit is on, and this is the only unit with a nonzero connection to the output unit.

Before leaving this example, we should note that there are values of the connection strengths that will do the job. One such set is shown in Figure 1E. The reader can check that this set produces the correct results for each of the four input patterns by using Equation 3.

Apparently, then, successful learning requires finding connection strengths that are not proportional to the correlations of activations of the units. How can this be done?

The Delta Rule

One answer that has occurred to many people over the years is the idea of using the difference between the desired, or *target*, activation and the obtained activation to drive learning. The idea is to adjust the strengths of the connections so that they will tend to reduce this *difference* or *error* measure. Because the rule is driven by differences, we have tended to call it the delta rule. Others have called it the Widrow-Hoff learning rule or the least mean square (LMS) rule (Widrow & Hoff, 1960); it is related to the perceptron convergence procedure of Rosenblatt (1959).

This learning rule, in its simplest form, can be written

$$\Delta w_{ij} = \epsilon e_i a_j \tag{4}$$

where e_i, the error for unit i, is given by

$$e_i = t_i - a_i, \tag{5}$$

the difference between the teaching input to unit i and its obtained activation.

To see how this rule works, let's use it to train the five-unit network in Figure 1C on the patterns in Figure 1D. The training regime is a little different here: For each pattern, we turn the input units on, then we see what effect they have on the output unit; its activation reflects the effects of the current connections in the network. (As before we assume the units are linear.) We compute the difference between the obtained output and the teaching input (Equation 5). Then, we adjust the strengths of the connections according to Equation 4. We will follow this procedure as we cycle through the four patterns several times, and look at the resulting strengths of the connections as we go. The network is started with initial weights of 0. The results of this process for the first cycle through all four patterns are shown in the first four rows of Figure 2.

The first time pattern 0 is presented, the response (that is, the obtained activation of the output unit) is 0, so the error is +1. This means that the changes in the weights are proportional to the activations of the input units. A value of 0.25 was used for the learning rate parameter, so each Δw is ± 0.25. These are added to the existing weights (which are 0), so the resulting weights are equal to these initial increments. When pattern 1 is presented, it happens to be uncorrelated with pattern 0, and so again the obtained output is 0. (The output is obtained by summing up the pairwise products of the inputs on the current trial with the weights obtained at the end of the preceding trial.) Again the error is +1, and since all the input units are on in this case, the change in the weight is +0.25 for each input. When these increments are added to the original weights, the result is a value of +0.5 for w_{04} and w_{24}, and 0 for the other weights. When the next pattern is presented, these weights produce an output of +1. The error is therefore −2, and so relatively larger Δw terms result. Even so, when the final pattern is presented, it produces an output of +1 as well. When the weights are adjusted to take this into account, the weight from input unit 0 is negative and the weight from unit 2 is positive; the other weights are 0. This completes the first sweep through the set of patterns. At this point,

```
Ep Pat    Input      Tgt Output Error       Delta w's              New values of w's

0  0  1 -1  1 -1|  1  0.00  1.00| 0.25-0.25 0.25-0.25 |  0.25-0.25 0.25-0.25
0  1  1  1  1  1|  1  0.00  1.00| 0.25 0.25 0.25 0.25 |  0.50 0.00 0.50 0.00
0  2  1  1  1 -1| -1  1.00 -2.00|-0.50-0.50-0.50 0.50 |  0.00-0.50 0.00 0.50
0  3  1 -1 -1  1| -1  1.00 -2.00|-0.50 0.50 0.50-0.50 | -0.50 0.00 0.50 0.00

                      tss: 10.00

1  0  1 -1  1 -1|  1  0.00  1.00| 0.25-0.25 0.25-0.25 | -0.25-0.25 0.75-0.25
1  1  1  1  1  1|  1  0.00  1.00| 0.25 0.25 0.25 0.25 |  0.00 0.00 1.00 0.00
1  2  1  1  1 -1| -1  1.00 -2.00|-0.50-0.50-0.50 0.50 | -0.50-0.50 0.50 0.50
1  3  1 -1 -1  1| -1  0.00 -1.00|-0.25 0.25 0.25-0.25 | -0.75-0.25 0.75 0.25

                      tss: 7.00

 ...

3  0  1 -1  1 -1|  1  0.25  0.75| 0.19-0.19 0.19-0.19 | -0.63-0.63 1.25 0.25
3  1  1  1  1  1|  1  0.25  0.75| 0.19 0.19 0.19 0.19 | -0.44-0.44 1.44 0.44
3  2  1  1  1 -1| -1  0.13 -1.13|-0.28-0.28-0.28 0.28 | -0.72-0.72 1.16 0.72
3  3  1 -1 -1  1| -1 -0.44 -0.56|-0.14 0.14 0.14-0.14 | -0.86-0.58 1.30 0.58

                      tss: 1.52

 ...

10 0  1 -1  1 -1|  1  0.90  0.10| 0.03-0.03 0.03-0.03 | -0.95-0.95 1.90 0.90
10 1  1  1  1  1|  1  0.90  0.10| 0.03 0.03 0.03 0.03 | -0.92-0.92 1.92 0.92
10 2  1  1  1 -1| -1 -0.85 -0.15|-0.04-0.04-0.04 0.04 | -0.96-0.96 1.89 0.96
10 3  1 -1 -1  1| -1 -0.92 -0.08|-0.02 0.02 0.02-0.02 | -0.98-0.94 1.91 0.94

                      tss: 0.05

 ...

20 0  1 -1  1 -1|  1  0.99  0.01| 0.00-0.00 0.00-0.00 | -1.00-1.00 1.99 0.99
20 1  1  1  1  1|  1  0.99  0.01| 0.00 0.00 0.00 0.00 | -1.00-1.00 2.00 1.00
20 2  1  1  1 -1| -1 -0.99 -0.01|-0.00-0.00-0.00 0.00 | -1.00-1.00 1.99 1.00
20 3  1 -1 -1  1| -1 -1.00 -0.00|-0.00 0.00 0.00-0.00 | -1.00-1.00 1.99 1.00

                      tss: 0.00
```

FIGURE 2. Learning with the delta rule. See text for explanation.

the values of the weights are far from perfect; if we froze them at these values, the network would produce 0 output to the first three patterns. It would produce the correct answer (an output of -1) only for the last pattern.

The correct set of weights is approached asymptotically if the training procedure is continued for several more sweeps through the set of patterns. Each of these sweeps, or *training epochs*, as we will call them henceforth, results in a set of weights that is closer to a perfect solution. To get a measure of the closeness of the approximation to a perfect solution, we can calculate an error measure for each pattern as that pattern is being processed. For each pattern, the error measure is the value of the error $(t-a)$ squared. This measure is then summed over all patterns to get a *total sum of squares* or *tss* measure. The resulting error measure, shown for each of the illustrated epochs in Figure 2, gets smaller over epochs, as do the changes in the strengths of the connections. The weights that result at the end of 20 epochs of training are very close to the perfect solution values. With more training, the weights converge to these values.

The error correcting learning rule, then, is much more powerful than the Hebb rule. In fact, it can be proven rather easily that the error-correcting rule will find a set of weights that drives the error as close to 0 as we want for each and every pattern in the training set, provided such a set of weights exists. Many proofs of this theorem have been given; a particularly clear one may be found in Minsky and Papert (1969).

The Linear Predictability Constraint

We have just noted that the delta rule will find a set of weights that solves a network learning problem, provided such a set of weights exists. What are the conditions under which such a set actually does exist?

Such a set of weights exists only if for each input-pattern–target-pair the target can be predicted from a weighted sum, or *linear combination*, of the activations of the input units. That is, the set of weights must satisfy

$$t_{ip} = \sum_j w_{ij} a_{jp} \tag{6}$$

for output unit i in all patterns p.

This constraint (which we called the *linear predictability constraint* in *PDP:17*) can be overcome by the use of hidden units, but hidden units cannot be trained using the delta rule as we have described it here because (by definition) there is no teacher for them. Procedures for training such units are discussed in Chapter 5.

Up to this point, we have considered the use of the Hebb rule and the delta rule for training connections coming into a single unit. We now consider how these learning rules produce the characteristics of *pattern associator* networks.

THE PATTERN ASSOCIATOR

In a pattern associator, there are two sets of units: input units and output units. There is also a matrix representing the connections from the input units to the output units. A pattern associator is really just an extension of the simple networks we have been considering up to now, in which the number of output units is greater than one and each input unit has a connection to each output unit. An example of an eight-unit by eight-unit pattern associator is shown in Figure 3.

The pattern associator is a device that learns associations between input patterns and output patterns. It is interesting because what it learns about

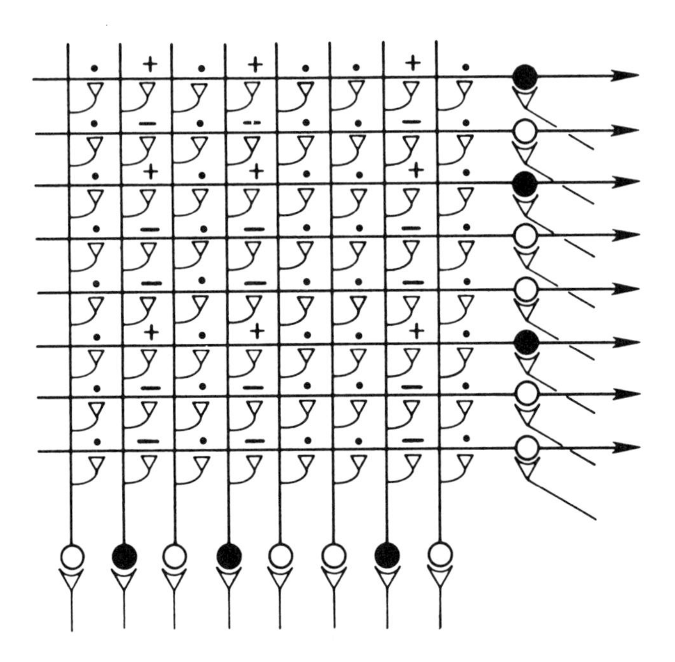

FIGURE 3. A schematic diagram of an eight-unit pattern associator. An input pattern, an output pattern, and values for the weights that will allow the input to produce the output are shown. (From *PDP:18*, p. 227.)

one pattern tends to generalize to other similar patterns. In what follows we will see how this property arises, first in the simplest possible pattern associator—a pattern associator consisting of linear units, trained by the Hebb rule.[1]

The Hebb Rule in Pattern Associator Models

To begin, let us consider the effects of training a network with a single learning trial l, involving an input pattern $\mathbf{i}_l$ and an output pattern $\mathbf{o}_l$. Assuming all the weights in the network are initially 0, we can express the value of each weight as

$$w_{ij} = \epsilon i_{jl} o_{il}. \tag{7}$$

Note that we are using the variable i_{jl} to stand for the activation of input unit j in input pattern $\mathbf{i}_l$, and we are using o_{il} to stand for the activation of output unit i in output pattern $\mathbf{o}_l$. Thus, each weight is just the product of

[1] Readers who wish to gain a better grasp on the mathematical basis of this class of models may find it worthwhile to read *PDP:9*. An in-depth analysis of the delta rule in pattern associators is in *PDP:11*.

the activation of the input unit times the activation of the output unit in the learning trial l.

Now let us present a test input pattern, $\mathbf{i}_t$, and examine the resulting output pattern it produces. Since the units are linear, the activation of output unit i when tested with input pattern $\mathbf{i}_t$ is

$$o_{it} = \sum_j w_{ij} i_{jt}. \tag{8}$$

Substituting for w_{ij} from Equation 7 yields

$$o_{it} = \sum_j \epsilon i_{jl} o_{il} i_{jt} \tag{9}$$

Since we are summing with respect to j in this last equation, we can pull out ϵ and o_{il}:

$$o_{it} = \epsilon o_{il} \sum_j i_{jl} i_{jt}. \tag{10}$$

Equation 10 says that the output at the time of test will be proportional to the output at the time of learning times the sum of the elements of the input pattern at the time of learning, each multiplied by the corresponding element of the input pattern at the time of test.

This sum of products of corresponding elements is called the *dot product.* It is very important to our analysis because it expresses the *similarity* of the two patterns $\mathbf{i}_l$ and $\mathbf{i}_t$. It is worth noting that we have already encountered an expression similar to this one in Equation 2. In that case, though, the quantity was proportional to the correlation of the activations of two *units* across an ensemble of *patterns.* Here, it is proportional to the correlation of two *patterns* across an ensemble of *units.* It is often convenient to normalize the dot product by taking out the effects of the number of elements in the vectors in question by dividing the dot product by the number of elements. We will call this quantity the *normalized dot product.* For patterns consisting of all +1s and −1s, it corresponds to the correlation between the two patterns. The normalized dot product has a value of 1 if the patterns are identical, a value of −1 if they are exactly opposite to each other, and a value of 0 if the elements of one vector are completely uncorrelated with the elements of the other.

We can rewrite Equation 10, then, replacing the summed quantity by the normalized dot product of input pattern $\mathbf{i}_l$ and input pattern $\mathbf{i}_t$, which we denote by $(\mathbf{i}_l \cdot \mathbf{i}_t)_n$:

$$o_{it} = k o_{il} (\mathbf{i}_l \cdot \mathbf{i}_t)_n \tag{11}$$

where $k = n\epsilon$ (n is the number of units). Since Equation 11 applies to all of the elements of the output pattern $\mathbf{o}_t$, we can write

$$\mathbf{o}_t = k \mathbf{o}_l (\mathbf{i}_l \cdot \mathbf{i}_t)_n. \tag{12}$$

This result is very basic to thinking in terms of patterns since it demonstrates that what is crucial for the performance of the network is the similarity relations among the input patterns—their correlations—rather than their specific properties considered as individuals.[2] Thus Equation 12 says that the output pattern produced by our network at test is a scaled version of the pattern stored on the learning trial. The magnitude of the pattern is proportional to the similarity of the learning and test patterns. In particular, if $k = 1$ and if the test pattern is identical to the training pattern, then the output at test will be identical to the output at learning.

An interesting special case occurs when the normalized dot product between the learned pattern and the test pattern is 0. In this case, the output is 0: There is no response whatever. Patterns that have this property are called *orthogonal* or *uncorrelated*; note that this is not the same as being opposite or *anticorrelated.*

To develop intuitions about orthogonality, you should compute the normalized dot products of each of the patterns *b*, *c*, *d*, and *e* below with pattern *a*:

a	+ + − −
b	+ − + −
c	+ − − +
d	+ + + +
e	− − + +

You will see that patterns *b*, *c*, and *d* are all orthogonal to pattern *a*; in fact, they are all orthogonal to each other. Pattern *e*, on the other hand, is not orthogonal to pattern *a*, but is anticorrelated with it. Interestingly, it forms an orthogonal set with patterns *b*, *c*, and *d*. When all the members of a set of patterns are orthogonal to each other, we call them an *orthogonal set.*

Now let us consider what happens when an entire ensemble of patterns is presented during learning. In the Hebbian learning situation, the set of weights resulting from an ensemble of patterns is just the sum of the sets of weights resulting from each individual pattern. That is, after learning trials on each of a set of input patterns $\mathbf{i}_l$, each paired with an output pattern $\mathbf{o}_l$, the value of each weight will be

$$w_{ij} = \epsilon \sum_l i_{jl} o_{il}. \tag{13}$$

Thus, the output produced by each test pattern is

$$\mathbf{o}_t = k \sum_l \mathbf{o}_l (\mathbf{i}_l \cdot \mathbf{i}_t)_n. \tag{14}$$

[2] Technically, performance depends on the similarity relations among the patterns and on their overall strength or magnitude. However, among vectors of equal strength (e.g., the vectors consisting of all $+1$s and -1s), only the similarity relations are important.

In words, the output of the network in response to input pattern t is the sum of the output patterns that occurred during learning, with each pattern's contribution weighted by the similarity of the corresponding input pattern to the test pattern. Three important facts follow from this:

1. If a test input pattern is orthogonal to all training input patterns, the output of the network will be 0; there will be no response to an input pattern that is completely orthogonal to all of the input patterns that occurred during learning.

2. If a test input pattern is similar to one of the learned input patterns and is uncorrelated with all the others, then the test output will be a scaled version of the output pattern that was paired with the similar input pattern during learning. The magnitude of the output will be proportional to the similarity of the test input pattern to the learned input pattern.

3. For other test input patterns, the output will always be a blend of the training outputs, with the contribution of each output pattern weighted by the similarity of the corresponding input pattern to the test input pattern.

In the exercises, we will see how these properties lead to several desirable features of pattern associator networks, particularly their ability to generalize based on similarity between test patterns and patterns presented during training.

These properties also reflect the limitations of the Hebbian learning rule; when the input patterns used in training the network do not form an orthogonal set, it is not in general possible to avoid contamination, or "cross-talk," between the response that is appropriate to one pattern and the response that occurs to the others. This accounts for the failure of Hebbian learning with the second set of training patterns considered in Figure 1. The reader can check that the input patterns we used in our first training example in Figure 1 (which was successful) were orthogonal but that the patterns used in the second example were not orthogonal.

The Delta Rule in Pattern Associator Models

Once again, the delta rule allows us to overcome the orthogonality limitation imposed by the Hebb rule. For the pattern associator case, the delta rule for a particular input-target pair $\mathbf{i}_l, \mathbf{t}_l$ is

$$\Delta w_{ij} = \epsilon (t_{il} - o_{il}) i_{jl}. \tag{15}$$

Therefore the weights that result from an ensemble of learning pairs indexed by l can be written:

$$w_{ij} = \epsilon \sum_l (t_{il} - o_{il}) i_{jl}. \tag{16}$$

It is interesting to compare this to the Hebb rule. Consider first the case where each of the learned patterns is orthogonal to every other one and is presented exactly once during learning. Then $\mathbf{o}_l$ will be $\mathbf{0}$ (a vector of all zeros) for all learned patterns l, and the above formula reduces to

$$w_{ij} = \epsilon \sum_l t_{il} i_{jl}. \tag{17}$$

In this case, the delta rule produces the same results as the Hebb rule; the teaching input simply replaces the output pattern from Equation 13. As long as the patterns remain orthogonal to each other, there will be no cross-talk between patterns. Learning will proceed independently for each pattern. There is one difference, however. If we continue learning beyond a single epoch, the delta rule will stop learning when the weights are such that they allow the network to produce the target patterns exactly. In the Hebb rule, the weights will grow linearly with each presentation of the set of patterns, getting stronger without bound.

In the case where the input patterns $\mathbf{i}_l$ are not orthogonal, the results of the two learning procedures are more distinct. In this case, though, we can observe the following interesting fact: We can read Equation 15 as indicating that the change in the weights that occurs on a learning trial is storing an association of the input pattern with the *error* pattern; that is, we are adding to each weight an increment that can be thought of as an association between the *error* for the output unit and the activation of the input unit. To see the implications of this, let's examine the effects of a learning trial with input pattern $\mathbf{i}_l$ paired with output pattern $\mathbf{t}_l$ on the output produced by test pattern $\mathbf{i}_t$. The effect of the change in the weights due to this learning trial (as given by Equation 15) will be to change the output of some output unit i by an amount proportional to the error that occurred for that unit on the learning trial, e_i, times the dot product of the learned pattern with the test pattern:

$$\Delta o_{it} = k e_{il} (\mathbf{i}_l \cdot \mathbf{i}_t)_n .$$

Here k is again equal to ϵ times the number of input units n. In vector notation, the change in the output pattern $\mathbf{o}_t$ can be expressed as

$$\Delta\, \mathbf{o}_t = k \mathbf{e}_l (\mathbf{i}_l \cdot \mathbf{i}_t)_n$$

Thus, the change in the output pattern at test is proportional to the error vector times the normalized dot product of the input pattern that occurred during learning and the input pattern that occurred during test. Two facts follow from this:

1. If the input on the learning trial is identical to the input on the test trial so that the normalized dot product is 1.0 and if $k = 1.0$, then the change in the output pattern will be exactly equal to the error pattern. Since the error pattern is equal to the difference between the target and the obtained output on the learning trial, this amounts to one trial learning of the desired association between the input pattern on the training trial and the target on this trial.

2. However, if $\mathbf{i}_t$ is different from $\mathbf{i}_l$ but not completely different so that $(\mathbf{i}_l \cdot \mathbf{i}_t)_n$ is not equal to either 1 or 0, then the output produced by $\mathbf{i}_t$ will be affected by the learning trial. The magnitude of the effect will be proportional to the magnitude of $(\mathbf{i}_l \cdot \mathbf{i}_t)_n$.

The second effect—the transfer from learning one pattern to performance on another—may be either beneficial or interfering. Importantly, for patterns of all +1s and −1s, the transfer is always less than the effect on the pattern used on the learning trial itself, since the normalized dot product of two different patterns must be less than the normalized dot product of a pattern with itself. This fact plays a role in several proofs concerning the convergence of the delta rule learning procedure (see Kohonen, 1977, and *PDP:11* for further discussion).

The Linear Predictability Constraint Again

Earlier we considered the linear predictability constraint for training a single output unit. Since the pattern associator can be viewed as a collection of several different output units, the constraint applies to each unit in the pattern associator. Thus, to master a set of patterns there must exist a set of weights w_{ij} such that

$$t_{ip} = \sum_j w_{ij} i_{jp} \tag{18}$$

for all output units i for all target-input pattern pairs p.

Another way of putting this set of constraints that is appropriate for the pattern associator is as follows: An arbitrary output pattern $\mathbf{o}_p$ can be correctly associated with a particular input pattern $\mathbf{i}_p$ without ruining associations between other input-output pairs, only if $\mathbf{i}_p$ *cannot* be written as a *linear combination* of the other input patterns. A pattern that cannot be written as a linear combination of a set of other patterns is said to be *linearly independent* from these other patterns. When all the members of a set of patterns are linearly independent, we say they form a *linearly independent set.* To ensure that arbitrary associations to each of a set of input patterns can be learned, the input patterns must form a linearly independent set.

It is worth noting that the linear independence constraint is primarily a constraint on the similarity relations among input patterns. If we consider the input patterns to be representations of environmental inputs, then whether a set of weights exists that allows us to associate arbitrary responses with each environmental input depends on the way in which these environmental inputs are represented as patterns of activation over a set of input units inside the system. As long as we already have a way of representing a set of environmental inputs so that they are linearly independent, the delta rule will be able to associate any arbitrary responses with these environmental inputs.

Although this is a serious constraint, it is worth noting that there are cases in which the response that we need to make to one input pattern can be predictable from the responses that we make to other patterns with which they overlap. In these cases, the fact that the pattern associator produces a response that is a combination of the responses to other patterns allows it to produce very efficient, often rule-like solutions to the problem of mapping each of a set of input patterns to the appropriate response. We will examine this property of pattern associators in the exercises.

Nonlinear Pattern Associators

Not all pattern associator models that have been studied in the literature make use of the linear activation assumptions we have been using in this analysis. Several different kinds of nonlinear pattern associators (i.e., associators in which the output units have nonlinear activation functions) fall within the general class of pattern associator models. These nonlinearities have effects on performance, but the basic principles that we have observed here are preserved even when these nonlinearities are in place. In particular:

1. Orthogonal inputs are mutually transparent.

2. The learning process converges with the delta rule as long as there is a set of weights that will solve the learning problem; the nonlinearities that have been tried tend not to have much effect on which sets of patterns can be learned and which cannot.

3. What is learned about one pattern tends to transfer to others.

THE FAMILY OF PATTERN ASSOCIATOR MODELS

With the above as background, we turn to a brief specification of several members of the class of pattern associator models that are available through the **pa** program. These are all variants on the pattern associator theme. Each model consists of a set of input units and a set of output units. The activations of the input units are *clamped* by externally supplied input patterns. The activations of the output units are determined in a single two-phase processing cycle. First, the net input to each output unit is computed. This is the sum of the activations of the input units times the corresponding weights, plus an optional bias term associated with the output unit:

$$net_i = \sum_j w_{ij} + bias_i. \tag{19}$$

Activation Functions

After computing the net input to each output unit, the activation of the output unit is then determined according to an activation function. Several variants are available:

- *Linear.* Here the activation of output unit i is simply equal to the net input.

- *Linear threshold.* In this variant, each of the output units is a *linear threshold unit*; that is, its activation is set to 1 if its net input exceeds 0 and is set to 0 otherwise. Units of this kind were used by Rosenblatt in his work on the perceptron (1959).

- *Stochastic.* This is the activation function used in *PDP:18* and *PDP:19*. Here, the output is set to 1, with a probability p given by the logistic function:

  $$p(o_i{=}1) = \frac{1}{1 + e^{-net_i/T}} \tag{20}$$

  This is the same activation function used in Boltzmann machines.

- *Continuous sigmoid.* In this variant, each of the output units takes on an activation that is nonlinearly related to its input according to the logistic function:

  $$o_i = \frac{1}{1 + e^{-net_i/T}} \tag{21}$$

Note that this is a continuous function that transforms net inputs between $+\infty$ and $-\infty$ into real numbers between 0 and 1. This is the activation function used in the back propagation networks we will study in Chapter 5.

Learning Assumptions

Two different learning rules are available in the **pa** program:

- *The Hebb rule.* Hebbian learning in the pattern associator model works as follows. Activations of input units are clamped based on an externally supplied input pattern, and activations of the output units are clamped to the values given by some externally supplied target pattern. Learning then occurs by adjusting the strengths of the connections according to the Hebbian rule:

$$\Delta w_{ij} = \epsilon o_i i_j . \tag{22}$$

- *The delta rule.* Error-correcting learning in the pattern associator model works as follows. Activations of input units are clamped to values determined by an externally supplied input pattern, and activations of the output units are calculated as described earlier. The difference between the obtained activation of the output units and the target activation, as specified in an externally supplied target pattern, is then used in changing the weights according to the following formula:

$$\Delta w_{ij} = \epsilon (t_i - o_i) i_j . \tag{23}$$

The Environment and the Training Epoch

In the pattern associator models, there is a notion of an *environment* of pattern pairs. Each pair consists of an input pattern and a corresponding output pattern. A training *epoch* consists of one learning trial on each pattern pair in the environment. On each trial, the input is presented, the corresponding output is computed, and the weights are updated. Patterns may be presented in fixed sequential order or in permuted order within each epoch.

Performance Measures

After processing each pattern, several measures of the output that is produced and its relation to the target are computed. One of these is the normalized dot product of the output pattern with the target. This measure is called the *ndp*. We have already described this measure quantitatively; here we note that it gives a kind of combined indication of the similarity of two patterns and their magnitudes. In the cases where this measure is most useful—where the target is a pattern of +1s and −1s—the magnitude of the target is fixed and the normalized dot product varies with the similarity of the output to the target and the magnitude of the output itself. To unconfound these factors, we provide two further measures: the *normalized vector length*, or *nvl*, of the output vector and the *vector correlation*, or *vcor*, of the output vector with the target vector. The *nvl* measures the magnitude of the output vector, normalizing for the number of elements in the vector. It has a value of 1.0 for vectors consisting of all +1s and −1s. The *vcor* measures the similarity of the vectors independent of their length; it has a value of 1.0 for vectors that are perfectly correlated, 0.0 for orthogonal vectors, and −1.0 for anticorrelated vectors.

Quantitative definitions of vector length and vector correlation are given in *PDP:9* (pp. 376-379).[3] For our purposes, it suffices to note the relation of these three measures. When the target pattern consists of +1s and −1s, the normalized dot product of the output pattern and the target pattern is equal to the normalized vector length of the output pattern times the vector correlation of the output pattern and the target:

$$ndp = nvl \times vcor. \tag{24}$$

In addition to these measures, we also compute the *pattern sum of squares* or *pss* and the *total sum of squares* or *tss*. The *pss* is the sum over all output units of the squared error, where the error for each output unit is the difference between the target and the obtained activation of the unit. This quantity is computed for each pattern processed. The *tss* is just the sum over the *pss* values computed for each pattern in the training set. These measures are not very meaningful when learning occurs by the Hebb rule, but they are meaningful when learning occurs by the delta rule.

---

[3] The normalization of vector length is not discussed in *PDP:9*. To compute the normalized vector length, one first divides the magnitude of each element by the square root of the number of elements in the vector, then computes the length of the resulting vector according to the formula given in *PDP:9*, p. 376.

IMPLEMENTATION

The **pa** program implements the pattern associator models in a very straightforward way. The program is initialized by defining a network, as in previous chapters. A PA network is assumed to consist of some number of input units (*ninputs*) and some number of output units (*noutputs*). Connections are allowed from input units to output units only. Each output unit may also have a bias term, which is treated as though it were a weight from a unit whose activation was set to 1.0, as in previous chapters. The network specification file defines the number of input units and output units, as well as the total number of units, and indicates which connections and bias terms to output units exist and which are modifiable. Usually, each input unit has a modifiable connection to each output unit, and all of these are initialized to 0, but other possibilities may be specified in the network specification file. It is also generally necessary to read in a file specifying the set of pattern pairs that make up the environment of the model.

Once the program is initialized, learning occurs through calls to a routine called *train*. This routine carries out *nepochs* of training, either in sequential order (if the routine is called by entering the *strain* command) or in permuted order (if the routine is called by entering the *ptrain* command). The routine exits if the total sum of squares measure, *tss*, is less than some criterion value, *ecrit*. Here is the *train* routine:

```
train(c) char c; {
/* c = 's' for strain, 'p' if ptrain */

  for (t = 0; t < nepochs; t++) {
    tss = 0.0;
    epochno++;

/* make a list of pattern numbers */
    for (i = 0; i < npatterns; i++)
      used[i] = i;

/* if ptrain, permute list */
    if (c == 'p') {
      for (i = 0; i < npatterns; i++) {
        npat = rnd() * (npatterns - i) + i;
        old = used[i];
        used[i] = used[npat];
        used[npat] = old;
      }
    }
    for (i = 0; i < npatterns; i++) {

/* set the pattern number, then do trial */
      patno = used[i];
```

```
        trial();
        if (lflag) change_weights();
      }
      if (tss < ecrit)
        return;
    }
}
```

The *trial* routine runs each individual trial. It calls four other routines: one that sets the input and the target patterns, one that computes the activations of the output units from the activations of the input units, one that computes the error measure, and one that computes the various summary statistics:

```
trial() {

  setup_pattern();
  compute_output();
  compute_error();
  sumstats();
}
```

Below we show the *compute_output* and the *compute_error* routines. First, *compute_output*:

```
compute_output() {

  for (i = ninputs; i < nunits; i++) {

/* accumulate net input to each output unit */
    netinput[i] = bias[i];
    for (j = 0; j < ninputs; j++) {
      netinput[i] += activation[j]*weight[i][j];
    }

/* set activation based on netinput */
    if (linear) {
      activation[i] = netinput[i];
    }
    else if (lt) {
      if (netinput[i] > 0)
        activation[i] = 1.0;
      else
        activation[i] = 0.0;
    }
    else if (cs) {
      activation[i] =  logistic(netinput[i]);
    }
```

```
    else { /* default, stochastic mode */
      activation[i] =
        probability(logistic(netinput[i]));
    }
  }
}
```

The choice of activation function is indicated by the values of mode variables called *linear*, *lt*, and *cs*. The variable *lt* stands for linear threshold, and *cs* stands for continuous sigmoid. The default case, in which the output is stochastic, involves a call to the *logistic* function shown in Chapter 3. This function returns a number between 0 and 1. The result is then used to set the activation of the unit to 0 or 1 based on the *probability* function, which returns a 1 with probability equal to its argument.

The *compute_error* function is exceptionally simple for the **pa** program:

```
compute_error() {

/* use i for output units, t for target elements */
  for (i = ninputs, t = 0; i < nunits; t++, i++) {
    error[i] = target[t] - activation[i];
  }
}
```

Note that when the targets and the activations of the output units are both specified in terms of 0s and 1s, the error will be 0, 1, or -1.

If learning is enabled (as it is by default in the program, as indicated by the value of the *lflag* variable), the *train* command calls the *change_weights* routine, which actually carries out the learning:

```
change_weights() {

  if (hebb) {
    for (i = ninputs; i < nunits; i++) {
      for (j = 0; j < ninputs; j++) {
         weight[i][j] +=
           epsilon[i][j]*target[i]*activation[j];
      }
      bias[i] += bepsilon[i]*target[i];
    }
  }

  else { /* delta rule, by default */
    for (i = ninputs; i < nunits; i++) {
      for (j = 0; j < ninputs; j++) {
        weight[i][j] +=
          epsilon[i][j]*error[i]*activation[j];
      }
```

```
            bias[i] += bepsilon[i]*error[i];
        }
    }
}
```

The matrix *epsilon[i][j]* contains modifiability parameters for each connection in the network. Generally, modifiable connections are set to a value equal to the parameter *lrate*; unmodifiable connections are set to 0. Similarly, the array *bepsilon[i]* has modifiability parameters for the bias terms for each output unit.

Note that for Hebbian learning, we use the target pattern directly in the learning rule, since this is mathematically equivalent to clamping the activations of the output units to equal the target pattern and then using these activations.

RUNNING THE PROGRAM

The **pa** program is used much like the other programs we have described in earlier chapters. The main things that are new for this program are the *strain* and *ptrain* commands for training pattern associator networks.

There are also changes to the *test* command, and there is a new command called *tall*. The *test* command allows you to test the network's response to a particular input pattern, either one in the program's list of pattern pairs or one entered directly from the keyboard; *tall* allows you to test the network's response to all of the patterns in the list of pattern pairs with learning turned off so as not to change the weights while testing. As in the **cs** program, commands *newstart* and *reset* are both available as alternative methods for reinitializing the programs. Recall that *reset* reinitializes the random number generator with the same seed used the last time the program was initialized, whereas *newstart* seeds the random number generator with a new random seed. Although there can be some randomness in **pa**, the problem of local minima does not arise and different random sequences will generally produce qualitatively similar results, so there is little reason to use *reset* as opposed to *newstart*. There are several *mode* variables in **pa**: *linear*, *lt* (for linear threshold), and *cs* (for continuous sigmoid) that determine the activation rule that is in force. The default is stochastic, which is used when *linear*, *lt*, and *cs* are all 0. Another mode variable, *hebb*, determines whether the learning rule is the Hebb rule (*hebb*=1) or the delta rule (which is the default). The new control variables include *nepochs*, the number of training epochs run when the *strain* and *ptrain* commands are entered, and *ecrit*, the criterion value for the error measure. The *lflag* variable allows direct control over whether learning is on, and the *stepsize* variable now has two new values, which are *pattern* and *epoch*. When *stepsize* is set to *pattern*, the screen is updated after each pattern is processed (i.e., at the end of each trial). When set to *epoch*, the screen is updated after each epoch of processing.

There are only three parameters in **pa**. Most important is *lrate*, which is equivalent to the parameter ϵ from the "Background" section. The other two parameters are *noise*, which determines the amount of random variability added to elements of input and target patterns, and *temp*, used as the denominator of the logistic function to scale net inputs in *cs* and the default stochastic mode. There are also several new performance measures: the normalized dot product, *ndp*; the normalized vector length measure, *nvl*; the vector correlation measure, *vcor*; the pattern sum of squares, *pss*; and the total sum of squares, *tss*.

New or Altered Commands

Here follows a more detailed description of the new or altered commands available in **pa**.

newstart
: Seeds the random number with a new random seed, and then returns the program to its initial state before any learning occurred. That is, sets all weights to 0, and sets *nepochs* to 0. Also clears activations and updates the display.

ptrain
: Permuted training command. Runs *nepochs* training epochs, presenting each pattern pair in the pattern list once in each epoch. Order of patterns is rerandomized for each epoch. Returns early if interrupted or if *tss* is less than *ecrit*. (During training, individual elements of each input and target pattern may be randomly distorted by the model by adding random noise, if desired. See the entry for the *noise* parameter in the variable list.)

reset
: Same as *newstart*, but reseeds the random number generator with the same seed that was used last time the network was initialized.

strain
: Sequential training command. Same as *ptrain*, except that pattern pairs are presented in the same, fixed order in each epoch. The order is simply the order in which the pattern pairs are encountered in the list.

tall
: Temporarily turns learning off and tests all patterns in sequential order. Temporarily sets *single* to 1 and sets *stepsize* to *pattern* so that the program updates the screen and pauses after each pattern is tested.

test

Allows testing of the response to an individually specified input pattern against an individually specified target. Prompts first for the input pattern, then for the target. Prompt for input is

input (#N, ?N, E for enter):

If the user types

#N Tests with the *N*th input pattern on the list (*N* is an integer).

#name Tests with the pattern named *name* (*name* is a character string).

?N Tests with a distorted version of pattern *N*. The pattern is distorted by adding uniformly distributed noise to the stored pattern with range as specified by the value of the *noise* parameter.

?name Tests with a distorted version of pattern *name*.

E Prompts for a specific pattern to test. The pattern consists of a sequence of floating-point numbers or "+", "–", and "." characters indicating entries for the pattern. Entries must be separated by spaces or *return* and followed by an extra *return* or *end*.

The prompt for the target is analogous; the user can specify a particular target to use in the same way the input pattern is specified. In this case, #0 indicates target pattern 0.

get/ patterns

Prompts for a file containing a set of pattern *pairs*. Each pair consists of an input pattern (called *ipattern*) and a target pattern (called *tpattern*). Each specification begins with a name, followed by *ninputs* entries specifying the elements of the input pattern and *noutputs* entries specifying the elements of the corresponding target. Entries may be floating-point numbers or "+" (stands for 1.0), "–" (stands for −1.0) or "." (stands for 0.0). Pattern pairs are numbered internally by the program in the order encountered starting from 0.

Variables

All of the new or altered variables used in the **pa** program are given in the following list. They are all accessible using the *set* and *exam* commands.

ecrit
: Error criterion for stopping training. If the *tss* at the end of an epoch of training is less than *ecrit*, training stops.

lflag
: Learning flag; normally set to 1. When nonzero, learning is enabled. Learning is automatically disabled during the *tall* or *test* commands.

nepochs
: Number of training epochs run by each call to the *strain* and *ptrain* commands.

stepsize
: Step size for updating the screen and for pausing if *single* is set to 1 during *strain* and *ptrain*. Values allowed are *cycle*, *pattern*, *epoch*, and *nepochs*. If the value is *cycle*, the screen is updated after processing each pattern and then updated again (if *lflag* is set to 1) after the weights are changed. If the value is *pattern*, the screen is only updated after the weights are changed. If the value is *epoch*, the screen is updated at the end of each epoch; if the value is *nepochs*, the screen is updated only when the *strain* and *ptrain* commands return.

config/ ninputs
: Number of input units. This is crucial both for defining the network and for reading in pattern pairs.

config/ noutputs
: Number of output units. Also crucial for defining the network and reading pattern pairs.

config/ nunits
: Number of units; should equal the sum of *ninputs* plus *noutputs*.

env/ ipattern
: Input pattern array. Prompts for an input pattern number or name, then for an element number, and then displays the value of the specified element in the specified pattern.

env/ npatterns
: Number of input-target pattern pairs in the program's pattern list. This variable is set implicitly in reading in the input patterns.

env/ pname
: Pattern name vector. Prompts for a pattern number and gives its name.

env/ tpattern
: Target pattern array; analogous to input pattern array. Prompts for a target pattern number or name and then for an element number. Element 0 is the first element of each target pattern.

mode/ cs
: Flag variable (standing for "continuous sigmoid"). When nonzero, each output unit takes on an activation based on the logistic function of its net input.

mode/ hebb

Flag variable. When nonzero, the Hebb rule is used in updating the weights; otherwise, the delta rule is used.

mode/ linear

Flag variable. When nonzero, a linear activation function is used to set the activations of the output units on the basis of the input units. If this and mode variables *lt* and *cs* are 0, the default stochastic activation function is used.

mode/ lt

Flag variable (standing for "linear threshold"). When nonzero, each output unit takes on an activation based on the linear threshold function so that its activation is set to 1 if its input is greater than 0; otherwise, its activation is set to 0.

param/ lrate

The learning rate parameter. Scales the size of the changes made to the weights. Generally, if there are n input units, the learning rate should be less than or equal to $1/n$.

param/ noise

Range of the random distortion added to each input and target pattern specification value during training and testing. The value added is uniformly distributed in the interval [−*noise*, +*noise*]. See the *strain*, *ptrain*, and *test* commands.

param/ temp

Denominator used in the logistic function to scale net inputs in both *cs* (continuous sigmoid) and default (stochastic) modes. Generally, *temp* can be set to 1, except in the simulations of the rule of 78 (see Ex. 4.4). Note that there is only one cycle of processing in **pa**, so there is no annealing.

state/ cpname

Name of the current pattern, as given in the pattern file.

state/ epochno

Number of the current epoch; updated at the beginning of each epoch.

state/ error

Vector of errors, or differences between the current target pattern and the current pattern of activation over the output units.

state/ input

Vector of activations of the input units in the network, based on the current input pattern (subject to the effects of noise).

state/ ndp

Normalized dot product of the obtained activation vector over the output units and the target vector.

state/ netinput

Vector of net inputs to each output unit.

state/ nvl
Normalized length of the obtained activation vector over the output units.

state/ output
Vector of activations of output units in the network.

state/ patno
The number of the current pattern, updated at the beginning of processing the pattern. Note that this is the index of the pattern on the program's pattern list; when *ptrain* is used, it is not the same as the pattern's position within the random training sequence in force for a particular epoch.

state/ pss
Pattern sum of squares, equal to the sum over all output units of the squared difference between the target for each unit and the obtained activation of the unit.

state/ target
Vector of target values for output units, based on the current target pattern, subject to effects of noise.

state/ tss
Total sum of squares, equal to the sum of all patterns so far presented during the current epoch of the pattern sum of squares.

state/ vcor
Vector correlation of the obtained activation vector over the output units and the target vector.

OVERVIEW OF EXERCISES

In these exercises, we will study several basic properties of pattern associator networks, starting with their tendency to generalize what they have learned to do with one input pattern to other similar patterns; we will explore the role of similarity and the learning of responses to unseen prototypes. These first studies will be done using a completely linear Hebbian pattern associator. Then, we will shift to the linear delta rule associator of the kind studied by Kohonen (1977) and analyzed in *PDP:11*. We will study what these models can and cannot learn and how they can be used to learn to get the best estimate of the correct output pattern, given noisy input and outputs. Finally, we will examine the acquisition of a rule and an exception to the rule in a nonlinear (stochastic) pattern associator.

Ex. 4.1. Generalization and Similarity With Hebbian Learning

In this exercise, you will train a linear Hebbian pattern associator on a single input-output pattern pair, and study how its output, after training, is affected by the similarity of the input pattern used at test to the input pattern used during training.

After you have created a working directory for the **pa** program and all of the associated files, you can start up this exercise by entering the command:

pa 8x8.tem lin.str

The files *8x8.tem* and *lin.str* set up the network to be a linear Hebbian pattern associator with eight input units and eight output units, starting with initial weights that are all 0. The *.str* file sets the value of the learning rate parameter to 0.125, which is equal to 1 divided by the number of units. With this value, the Hebb rule will learn an association between a single input pattern consisting of all +1s and −1s and any desired output pattern perfectly in one trial.

The file *one.pat* contains a single pattern (or, more exactly, a single input-output pattern pair) to use for training the associator. Both the input pattern and the output pattern are eight-element vectors of +1s and −1s. Load this pattern file:

get pat one.pat

When it is loaded, the single input and target pattern will appear in the upper right corner of the display area.

Now you can train the network on this first pattern pair for one epoch. The *nepochs* variable has been set to 1 in the *lin.str* file, but before you start it might be best to set *single* to 1 so that you can watch things progress:

set single 1

Now, you can train the network by simply entering

strain

Because *stepsize* is set to *cycle* in the *.str* file, the program will present the first (and, in this case, only) input pattern, compute the output based on the current weights, and then display the input, output, and target patterns, as well as some summary statistics. If you have set *single* to 1, it will pause, with the display shown in Figure 4. At the top of the display you will see the **p to push /b to break /<cr> to continue:** prompt, under which is the top-level menu. Below this is the display area. In the upper left corner of the display area, you will see some summary information, including the current *epochno* (which is 1, indicating the first epoch of training; note that *epochno* changes at the beginning of each epoch and stays the same until the next epoch begins) and the current *patno* (which is 0, indicating that pattern 0 is being processed). Below these entries are the *ndp*, or normalized dot product, of the output obtained by the network with the target pattern; the *nvl*, or normalized vector length, of the obtained output pattern;

```
p to push/b to break/<cr> to continue:
disp/  exam/  get/  save/  set/  clear  do  log  newstart  ptrain  quit  reset
run  strain  tall  test

epochn      1                                          pname  ipattern  tpattern
cpname      a                                          a      11111111  11111111
ndp     0.0000
nvl     0.0000
vcor    0.0000
pss     8.0000
tss     8.0000                              out   tar

weights     0  0  0  0  0  0  0  0           0    100
            0  0  0  0  0  0  0  0           0    100
            0  0  0  0  0  0  0  0           0    100
            0  0  0  0  0  0  0  0           0    100
            0  0  0  0  0  0  0  0           0    100
            0  0  0  0  0  0  0  0           0    100
            0  0  0  0  0  0  0  0           0    100
            0  0  0  0  0  0  0  0           0    100

input     100100100100100100100100
```

FIGURE 4. Display layout for the first **pa** exercise while processing pattern a, before any learning has occurred.

and the *vcor*, or vector correlation, of the output with the target. All of these numbers are 0 because the weights are 0, so the input produces no output at all. Below these numbers are the *pss*, or pattern sum of squares, and the *tss*, or total sum of squares. They are the sum of squared differences between the target and the actual output patterns. The first is summed over all output units for the current pattern, and the second is summed over all patterns so far encountered within this epoch (they are, therefore, identical at this point).

Below these entries you will see the weight matrix on the left, with the input vector that was presented for processing below it and the output and target vectors to the right.

The display presents all of the network variables (inputs, weights, activations, etc.) in *hundredths*, and reverse video means negative, so that 1.0 displays as 100 and -1.0 displays as 100 in reverse video. The *ipattern* and *tpattern* arrays (upper right corner of the display area) show only one digit per pattern element because they are always $+1$ and -1 in this exercise. You will see that the input pattern shown below the weights matches the single input pattern shown under the label *ipattern* and that the target pattern shown to the right of the weights matches the single target pattern shown under *tpattern*. The output itself is all 0s.

If you enter *return* at this point, the target will first be clamped onto the output units, then the weights will be updated according to the Hebbian learning rule:

$$\Delta w_{ij} = (lrate) o_i i_j . \tag{25}$$

Enter *return* one more time to return from the first epoch of training.

Q.4.1.1. Explain the values of the weights in rows 1 and 2 (counting from 0). Note that the display does not indicate the third decimal place, which in this case is a 5; each weight has absolute value 0.125 at this point. Explain the values of the weights in column 7, the last column of the matrix.

Now, with just this one trial of learning, the network will have "mastered" this particular association, so that if you test it at this point, you will find that, given the learned input, it perfectly reproduces the target. You can test the network using the *test* command. Simply enter *test*, and respond to the prompts. The first prompt asks you to specify an input pattern. Your choices are *#N, ?N,* and *E*, where *N* stands for the numerical index of a pattern. If you enter *#0*, for example, the model will set the input to be *ipattern* 0. The *?N* option is used to specify a distortion of a particular pattern (we will consider this again in a moment); the *E* option allows you to enter a specific pattern of your own choosing. For your first test, enter *#0* for the input. Do the same for the target. The display will now show the actual output computed by the network.

In this particular case the display will not change much because in the previous display the output had been clamped to reflect the very target pattern that the network has now computed. The only thing that actually changes in the display are the *ndp*, *vcor*, and *nvl* fields; these will now reflect the normalized dot product and correlation of the computed output with the target and the normalized length of the output. They should all be equal to 1.0 at this point.

You are now ready to test the generalization performance of the network. To do this, you can enter specific input patterns of your own, instead of *#0*, for the input pattern requested by the test routine. The following exchange indicates the inputting of the pattern +−+−++++, followed by a specification that the "target" is still *tpattern* 0:

> **pa**: *test*
> **input (#N, ?N, E for enter)**: *E*
> **give input elements**: + − + − + + + + *end*
> **target (#N, ?N, E for enter)**: *#0*

Note that the elements of the input pattern must be separated by spaces. After this interchange, the network will test the specified input pattern, report the output and the target, and indicate the *ndp, vcor,* and *nvl* of the obtained output with the input.

Q.4.1.2. Try a number of different input patterns, testing each against the *#0* target. Observe the *ndp, vcor*, and *nvl* in each case. Relate the obtained output to the specifics of the weights and the input patterns used and to the discussion in the "Background" section about the test output we should get from a linear Hebbian associator, as a function of the normalized dot product of the input vector used at test and the input vector used during training. Include in your set of patterns one that is orthogonal to the training pattern and one that is perfectly anticorrelated with it, as well as one or two others with positive normalized dot products with the input pattern.

If you understand the results you have obtained in this exercise, you understand the basis of similarity-based generalization in one-layer associative networks. In the process, you should come to develop your intuitions about vector similarity and to clearly be able to distinguish uncorrelated patterns from anticorrelated ones.

Ex. 4.2. Orthogonality, Linear Independence, and Learning

This exercise will expose you to the limitation of a Hebbian learning scheme and show how this limitation can be overcome using the delta rule. For this exercise, you are to set up two different sets of training patterns: one in which all the input patterns form an orthogonal set and the other in which they form a linearly independent, but not orthogonal, set. For both cases, choose the output patterns so that they form an orthogonal set, then arbitrarily assign one of these output patterns to go with each input pattern. In both cases, use only three pattern pairs and make sure that both patterns in each pair are eight elements long. The pattern files you construct in each case should contain three lines formatted like the single line in the *one.pat* file:

```
first + - + - + - + -   + + - - + + - -
```

We provide sets of patterns that meet these conditions in the two files *ortho.pat* and *li.pat*. You may use these files if you wish, but you will be more certain of your understanding if you construct your own sets of patterns and make sure that they can be learned.

Q.4.2.1. For each set of patterns, do the following experiment. Read the patterns into the program using the *get/ patterns* command. Reset the network (this clears the weights to 0s). Then run one epoch of training using the Hebbian learning rule by entering the *strain* command. Following this, execute the *tall* command to test the model's performance. What happens with each pattern? Run three additional epochs of training (one at a time), following each with a *tall*. What happens? In what ways do things get better? In what ways do they stay the same? Why?

Q.4.2.2. Turn off *hebb* mode in the program (this puts the default, the delta rule, in place) and try the above experiment again. Describe the similarities and differences between the results and explain in terms of the differential characteristics of the Hebbian and delta rule learning schemes.

Q.4.2.3. With the linearly independent set of patterns, keep running training epochs using the delta rule until the *tss* measure drops below 0.05. Examine and explain the resulting weight matrix, contrasting it with the weight matrix obtained after one cycle of Hebbian learning with the same patterns. What are the similarities between the two matrices? What are the differences? Try to explain rather than just describe the differences.

For the next question, reinitialize your network and train the network with the pattern set in the file *li.pat*. (It makes no difference whether you use *reset* or *newstart* to reinitialize since randomness plays no role in this exercise.) Run one epoch of training at a time, and examine performance at the end of each epoch using the *tall* command.

Q.4.2.4. In *li.pat*, one of the input patterns is orthogonal to both of the others, which are partially correlated with each other. Which input-output pair is mastered first? Why?

Q.4.2.5. As the final exercise in this set, construct sets of two or more pattern pairs that cannot be effectively mastered, either by Hebbian or delta rule learning. Explain why they cannot be learned, and describe what happens when the network tries to learn them, both in terms of the course of learning and in terms of the weights that result.

Hints. We provide a set of impossible pattern pairs in the file *imposs.pat*, but it is preferable for you to try to construct your own. You will probably want to use a small value of the learning rate; this affects the size of the oscillations that you will probably observe in the weights. A learning rate of about 0.0125 is probably good. Keep running more training epochs until the *tss* at the end of each epoch stabilizes.

Ex. 4.3. Learning Central Tendencies

One of the positive features of associator models is their ability to filter out noise in their environments. In this exercise we invite you to explore this aspect of pattern associator networks. For this exercise, you will still be using linear units but with the delta rule and with a relatively small learning rate. You will also be introducing noise into your training patterns.

A start-up file that sets things up nicely for this exercise is provided, called *ct.str* (*ct* is for "central tendency"). This file sets the learning rate to 0.0125 and makes sure *hebb* mode is off. It also sets the *noise* variable to 0.5. This means that each element in each input pattern and in each target pattern will have its activation distorted by a random amount uniformly distributed between +0.5 and −0.5.

Start up the program with the command

pa 8x8.tem ct.str

Then load in a set of patterns (your orthogonal set from Ex. 4.2 or the patterns in *ortho.pat*). Then you can see how well the model can do at pulling out the "signals" from the "noise." The clearest way to see this is by studying the weights themselves and comparing them to the weights acquired with the same patterns without noise added.

Q.4.3.1. Compare learning of the three orthogonal patterns you used in Ex. 4.2, without noise, to the learning that occurs in this exercise, with noise added. Compare the weight matrix acquired after "noiseless" learning with the matrix that evolves given the noisy input-target pairs that occur in the current situation. Run about 60 epochs of training to get an impression of the evolution of the weights through the course of training and compare the results to what happens with errorless training patterns (and a higher learning rate). What effect does changing the learning rate have when there is noise? Try higher and lowers values.

Hints. You may find it useful to rerun the relevant part of Ex. 4.2 (Q.4.2.2) and to use the *save/ screen* command to store the weight matrices you obtain in the different runs. For longer runs, remember that you can set *nepochs* to a number larger than the default value of 1. Each time *strain* is entered, *nepochs* of training are run.

The results of this simulation are relevant to the theoretical analyses described in *PDP:11* and are very similar to those described under "central tendency learning" in *PDP:25*, where the effects of amnesia (taken as a reduction in connection strength) are considered.

Ex. 4.4. Lawful Behavior

We now turn to one of the principle characteristics of pattern associator models that has made us take interest in them: their ability to pick up regularities in a set of input-output pattern pairs. The ability of pattern associator models to do this is illustrated in the past-tense learning model, discussed in *PDP:18*. Here we provide the opportunity to explore this aspect of pattern associator models, using the example discussed in that chapter, namely, the *rule of 78* (see *PDP:18*, pp. 226-234). We briefly review this example here.

The rule of 78 is a simple rule we invented for the sake of illustration. The rule first defines a set of eight-element input patterns. In each input pattern, one of units 1, 2, and 3 must be on; one of units 4, 5, and 6 must be on; and one of units 7 and 8 must be on. For the sake of consistency with *PDP:18*, we adopt the convention for this example only of numbering units starting from 1. The rule of 78 also defines a mapping from input to output patterns. For each input pattern, the output pattern that goes with it is the same as the input pattern, except that if unit 7 is on in the input pattern, unit 8 is on in the output and vice versa. Table 1 shows this rule.

The rule of 78 defines 18 input-output pattern pairs. Eighteen *arbitrary* input-output pattern pairs would exceed the capacity of an eight-by-eight pattern associator, but as we shall see, the patterns that exemplify the rule of 78 can easily be learned by the network.

The version of the pattern associator used for this example follows the assumptions we adopted in *PDP:18* for the past-tense learning model. Input units are binary and are set to 1 or 0 according to the input pattern. The output units are binary, stochastic units and take on activation values of 0 or 1 with probability given by the logistic function:

$$p(act_i = 1) = \frac{1}{1 + e^{-net_i/T}} \tag{26}$$

TABLE 1

THE RULE OF 78

Input patterns consist of one active unit from each of the following sets:	(1 2 3) (4 5 6) (7 8)
The output pattern paired with a given input pattern consists of:	The same unit from (1 2 3) The same unit from (4 5 6) The other unit from (7 8)
Examples:	2 4 7 → 2 4 8 1 6 8 → 1 6 7 3 5 7 → 3 5 8
An exception:	1 4 7 → 1 4 7

(From *PDP:18*, p. 229.)

where T is equivalent to the *temp* parameter. Note that, although this function is the same as for the Boltzmann machine, the calculation of the output is only done once, as in other versions of the pattern associator; there is no annealing, so *temp* is just a scaling factor.

Learning occurs according to the delta rule, which in this case is equivalent to the perceptron convergence procedure because the units are binary. Thus, when an output unit should be on (target is 1) but is not (activation is 0), an increment of size *lrate* is added to the weight coming into that unit from each input unit that is on. When an output unit should be off (target is 0) but is not (activation is 1), an increment of size *lrate* is subtracted from the weight coming into that unit from each input unit that is on.

For this example, we follow *PDP:18* and use *temp* of 15. This means that the net input to a unit needs to be about 45 for it to come on with probability .95. Thus, learning will be gradual, even with an lrate of 2.0, which is what we use in these examples. Note, though, that you can speed things up by using a larger value of the learning rate. (The simulations that you will do here will not conform to the example in *PDP:18* in all details, since in that example an approximation to the logistic function was used. The basic features of the results are the same, however.)

To run this example, you will need to start up the **pa** program with the *78.tem* and *78.str* files:

pa 78.tem 78.str

The *78.str* file will read in the appropriate network specification file (in *8x8.net*) and the 18 patterns that exemplify the rule of 78, then display these on the screen to the right of the weight matrix. Since the units are binary, there is only a single digit of precision for both the input, output, and target units. Given the *lrate* of 2.0, the weights take on integer values divisible by 2. Their actual values are displayed in the cells of the weight matrix; they are not multiplied first by 100, as in the previous exercises.

You should now be ready to run the exercise. The variable *nepochs* is initialized to 10, so if you enter one of the training commands, 10 epochs of training will be run. We recommend using *ptrain* because it does not result in a consistent bias in the weights favoring the patterns later in the pattern list. The screen is updated once per pattern after the weights have been adjusted, so you should see the weights and the input, output, and target bits changing. The *pss* and *tss* (which in this case indicate the number of incorrect output bits) will also be displayed once per pattern.

Q.4.4.1. At the end of the 10th epoch, the *tss* should be in the vicinity of 20, or about one error per pattern; this means that the model is getting each output bit correct with a probability of about .9. Given the values of the weights and the fact that *temp* is set to 15, calculate the net input to the last output unit for the first two input patterns, and (using Figure 7 of Chapter 3 or your calculator) calculate the approximate probability that this last output unit will receive the correct activation in each of these two patterns. The probabilities should be in the upper .80s or low .90s.

At this point you should be able to see the solution to the rule of 78 patterns emerging. Generally, there are large positive weights between input units and corresponding output units, with unit 7 exciting unit 8 and unit 8 exciting unit 7. You'll also see rather large inhibitory weights from each input unit to each other unit within the same subgroup (i.e., 1, 2, and 3; 4, 5, and 6; and 7 and 8). Run another 20 or so epochs, and a subtler pattern will begin to emerge.

Q.4.4.2. Generally there will be slightly negative weights from input units to output units in other subgroups. See if you can understand why this happens. Note that this does not happen reliably for weights coming into output units 7 and 8. Your explanation should explain this too.

At this point, you have watched a simple PDP network learn to behave in accordance with a simple rule, using a simple, local learning scheme; that is, it adjusts the strength of each connection in response to its errors on each particular learning experience, and the result is a system that exhibits lawful behavior in the sense that it conforms to the rule.

For the next part of the exercise, you can explore the way in which this kind of pattern associator model captures the three-stage learning phenomenon exhibited by young children learning the past tense in the course of learning English as their first language. To briefly summarize this phenomenon: Early on, children know only a few words in the past tense. Many of these words happen to be exceptions, but at this point children tend to get these words correct. Later in development, children begin to use a much larger number of words in the past tense, and these are predominantly regular. At this stage, they tend to overregularize exceptions. Gradually, over the course of many years, these exceptions become less frequent, but adults have been known to say things like *ringed* or *taked*, and lower-frequency exceptions tend to lose their exceptionality (i.e., to become regularized) over time.

The 78 model can capture this pattern of results; it is interesting to see it do this and understand how and why this happens. For this part of the exercise, you will want to reset the weights, and read in the file *hf.pat*, which contains a exception pattern *(147 → 147)* and one regular pattern *(258 → 257)*. If we imagine that the early experience of the child consists mostly of exposure to high-frequency words, a large fraction of which are irregular (8 of the 10 most frequent verbs are irregular), this approximates the early experience the child might have with regular and irregular past-tense forms. If you run 20 epochs of training using *ptrain* with these two patterns, you will see a set of weights that allows the model to set each output bit correctly in each pattern about 85% of the time. At this point, you can read in the file *all.pat*, which contains these two pattern pairs, plus all of the other pairs that are consistent with the rule of 78. This file differs from the *78.pat* file only in that the input pattern *147* is associated with the "exceptional" output pattern *147* instead of what would be the "regular" corresponding pattern *148*. Save the screen displaying the weights that resulted from learning *hf.pat*. Then read in *all.pat* and run 10 more epochs.

Q.4.4.3. Given the weights that you see at this point, what is the network's most probable response to *147*? Can you explain why the network has lost the ability to produce *147* as its response to this input pattern? What has happened to the weights that were previously involved in producing *147* from *147*?

One way to think about what has happened in learning the *all.pat* stimuli is that the 17 regular patterns are driving the weights in one direction and the single exception pattern is fighting a lonely battle to try to drive the weights in a different direction, at least with respect to the activation of units 7 and 8. Since eight of the input patterns have unit 7 on and "want" output unit 8 to be on and unit 7 to be off and only one input pattern has input unit 7 on and wants output unit 7 on and output unit 8 off, it is hardly a fair fight.

If you run more epochs, though, you will find that the network eventually finds a compromise solution that satisfies all of the patterns.

Q.4.4.4. If you have the patience, run the model until it finds a set of weights that gets each output unit correct about 90% of the time for each input pattern (90% correct corresponds to a net input of +30 or so for units that should be on and −30 for units that should be off). Explain why it takes so long to get to this point.

Further Suggested Exercises

In the exercise just described, there was only one exception pattern, and when vocabulary size increased, the ratio of regular to exception patterns increased from 1:1 to 17:1. Pinker and Prince (1987) have shown that, in fact, as vocabulary size increases, the ratio of regular to exception verbs stays roughly constant at 1:1. One interesting exercise is to set up an analog of this situation. Start training the network with one regular and one exception pattern, then increase the "vocabulary" by introducing new regular patterns and new exceptions. Note that each exception should be idiosyncratic; if all the exceptions were consistent with each other, they would simply exemplify a different rule. You might try an exercise of this form, setting up your own correspondence rules, your own exceptions, and your own regime for training.

You can also explore other variants of the pattern associator with other kinds of learning problems. One thing you can do easily is see whether the model can learn to associate each of the individuals from the Jets and Sharks example in Chapter 2 with the appropriate gang (relying only on their properties, not their names; the files *jets.tem, jets.str, jets.net*, and *jets.pat* are available for this purpose). Also, you can play with the continuous sigmoid (or logistic) activation function by setting the *cs* mode flag to 1 (for the rule of 78 example, it is best to use the *8x8.tem* file since the weights have real values and to cut *lrate* to about 0.12 and *temp* to 1.0 so that the weights stay in range; remember weights and activations are displayed as hundredths in this template).

CHAPTER 5

Training Hidden Units: The Generalized Delta Rule

In this chapter, we introduce the back propagation learning procedure for learning internal representations. We begin by describing the history of the ideas and problems that make clear the need for back propagation. We then describe the procedure, focusing on the goal of helping the student gain a clear understanding of gradient descent learning and how it is used in training PDP networks. The exercises are constructed to allow the reader to explore the basic features of the back propagation paradigm. At the end of the chapter, there is a separate section on extensions of the basic paradigm, including three variants we call *cascaded* back propagation networks, *recurrent* networks, and *sequential* networks. Exercises are provided for each type of extension.

BACKGROUND

The pattern associator described in the previous chapter has been known since the late 1950s, when variants of what we have called the delta rule were first proposed. In one version, in which output units were linear threshold units, it was known as the perceptron (cf. Rosenblatt, 1959, 1962). In another version, in which the output units were purely linear, it was known as the LMS or least mean square associator (cf. Widrow & Hoff, 1960). Important theorems were proved about both of these versions. In the case of the perceptron, there was the so-called perceptron convergence theorem. In this theorem, the major paradigm is pattern classification. There is a set of binary input vectors, each of which can be said to belong to one of two classes. The system is to learn a set of connection strengths and a threshold value so that it can correctly classify each of the input vectors. The basic structure of the perceptron is illustrated in Figure 1. The perceptron learning procedure is the following: An input vector is presented to the system (i.e., the input units are given an activation of 1 if the corresponding value of the input vector is 1 and are given 0 otherwise). The net input to the output unit is computed: $net = \sum_i w_i i_i$. If *net* is greater than the threshold θ, the unit is turned on, otherwise it is turned off. Then the response is compared with the actual category of the input vector. If the vector was correctly categorized, then no change is made to the weights. If, however, the output unit turns on when the input vector is in category 0, then the weights and thresholds are modified as follows: The threshold is incremented by 1 (to make it less likely that the output unit will come on if the same vector were presented again). If input i_i is 0, no change is made in the weight w_i (that weight could not have contributed to its having turned on). However, if $i_i = 1$, then w_i is decremented by 1. In this way, the system will not be as likely to turn on the next time this input vector is presented. On the other hand, if the output unit does not come on when it is supposed to, the opposite changes are made. That is, the threshold is decremented, and those weights connecting the output units to input units that are on are incremented.

Mathematically, this amounts to the following: The output, o, is given by

$$o = \begin{cases} 1 & \text{if } net = \sum_i w_i i_i > \theta \\ 0 & \text{otherwise.} \end{cases}$$

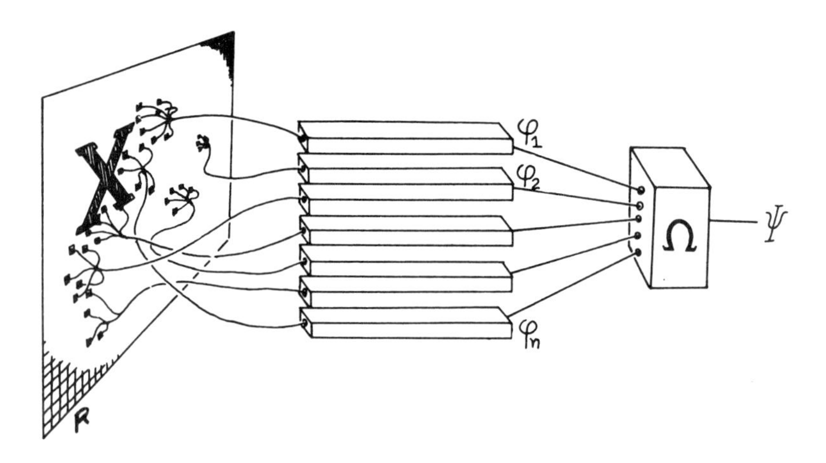

FIGURE 1. The one-layer perceptron analyzed by Minsky and Papert. (From *Perceptrons* by M. L. Minsky and S. Papert, 1969, Cambridge, MA: MIT Press. Copyright 1969 by MIT Press. Reprinted by permission.)

The change in the threshold, $\Delta\theta$, is given by

$$\Delta\theta = -(t_p - o_p) = -\delta_p$$

where p indexes the particular pattern being presented, t_p is the target value indicating the correct classification of that input pattern, and δ_p is the difference between the target and the actual output of the network. Finally, the changes in the weights, Δw_i, are given by

$$\Delta w_i = (t_p - o_p)i_{pi} = \delta_p i_{pi}.$$

The remarkable thing about this procedure is that, in spite of its simplicity, such a system is guaranteed to find a set of weights that correctly classifies the input vectors *if such a set of weights exists.* Moreover, since the learning procedure can be applied independently to each of a set of output units, the perceptron learning procedure will find the appropriate mapping from a set of input vectors onto a set of output vectors—*if such a mapping exists.* Unfortunately, as indicated in Chapter 4, such a mapping does not always exist, and this is the major problem for the perceptron learning procedure.

In their famous book *Perceptrons,* Minsky and Papert (1969) document the limitations of the perceptron. The simplest example of a function that cannot be computed by the perceptron is the exclusive-or (XOR), illustrated in Table 1. It should be clear enough why this problem is impossible. In order for a perceptron to solve this problem, the following four inequalities must be satisfied:

$$0 \times w_1 + 0 \times w_2 < \theta => 0 < \theta$$

$$0 \times w_1 + 1 \times w_2 > \theta => w_1 > \theta$$

$$1 \times w_1 + 0 \times w_2 > \theta => w_2 > \theta$$

$$1 \times w_1 + 1 \times w_2 < \theta => w_1 + w_2 < \theta$$

Obviously, we can't have both w_1 and w_2 greater than θ while their sum, $w_1 + w_2$, is less than θ. There is a simple geometric interpretation of the class of problems that can be solved by a perceptron: It is the class of

TABLE 1

Input Patterns		Output Patterns
00	→	0
01	→	1
10	→	1
11	→	0

(From *PDP:8*, p. 319)

linearly separable functions. This can easily be illustrated for two-dimensional problems such as XOR. Figure 2 shows a simple network with two inputs and a single output and illustrates three two-dimensional functions: the AND, the OR, and the XOR. The first two can be computed by the network; the third cannot. In these geometrical representations, the input patterns are represented as coordinates in space. In the case of a binary two-dimensional problem like XOR, these coordinates constitute the vertices of a square. The pattern 00 is represented at the lower left of the square, the pattern 10 as the lower right, and so on. The function to be computed is then represented by labeling each vertex with a 1 or 0 depending on which class the corresponding input pattern belongs to. The perceptron can solve any function in which a single line can be drawn through the space such that all of those labeled "0" on are one side of the line and those labeled "1" are on the other side. This can easily be done for AND and OR, but not for XOR. The line corresponds to the equation $i_1w_1+i_2w_2=\theta$. In three dimensions there is a plane, $i_1w_1+i_2w_2+i_3w_3=\theta$, that corresponds to the line. In higher dimensions there is a corresponding

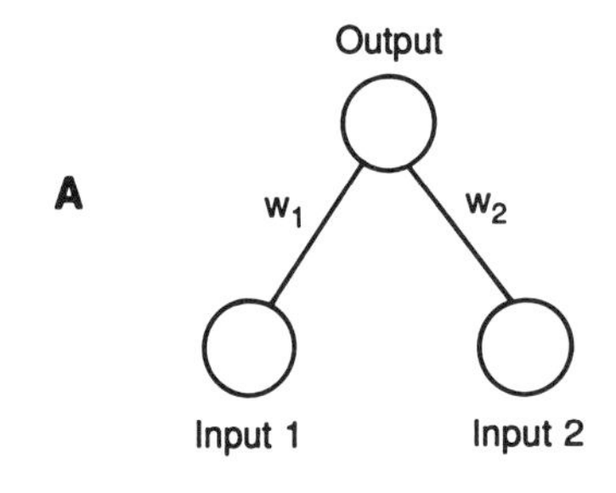

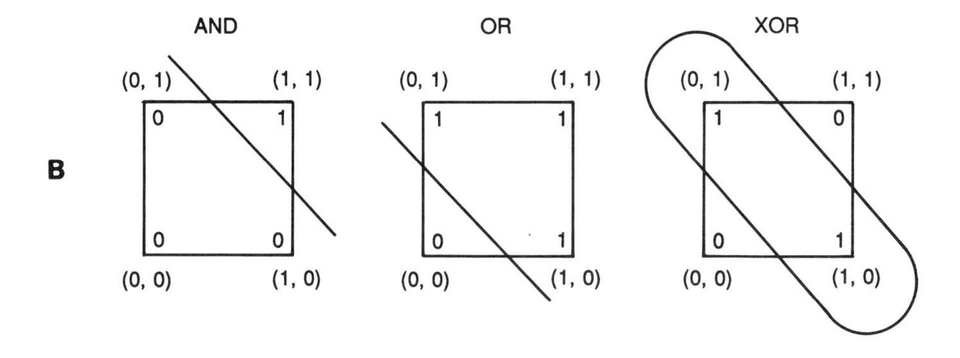

FIGURE 2. *A:* A simple network that can solve the two-dimensional AND and OR functions but cannot solve the XOR function. *B:* Geometric representations of the three problems. See text for details.

TABLE 2

Input Patterns		Output Patterns
000	→	0
010	→	1
100	→	1
111	→	0

(From *PDP:8*, p. 319.)

hyperplane, $\sum_{i=1}^{n} w_i i_i = \theta$. All functions for which there exists such a plane are called *linearly separable.* Now consider the function in Table 2 and illustrated in Figure 3. This is a three-dimensional problem in which the first two dimensions are identical to the XOR and the third dimension is the AND of the first two dimensions. (That is, the third dimension is 1 whenever both of the first two dimensions are 1, otherwise it is 0). Figure 3 shows how this problem can be represented in three dimensions. The figure also shows how the addition of the third dimension allows a plane to separate the patterns classified in category 0 from those in category 1. Thus, we see that the XOR is not solvable in two dimensions, but if we add the appropriate third dimension, that is, the appropriate *new feature,* the problem *is* solvable. Moreover, as indicated in Figure 4, if you allow a multilayered perceptron, it is possible to take the original two-dimensional problem and convert it into the appropriate three-dimensional problem so it can be solved. Indeed, as Minsky and Papert knew, it is always possible to convert any unsolvable problem into a solvable one in a multilayer perceptron. In the more general case of multilayer networks, we categorize units

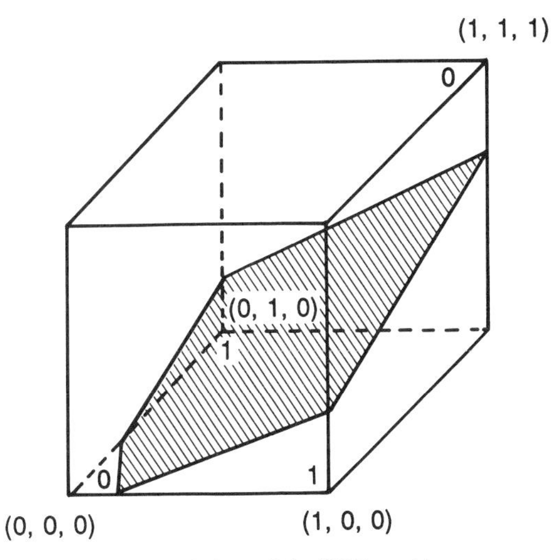

FIGURE 3. The three-dimensional solution of the XOR problem.

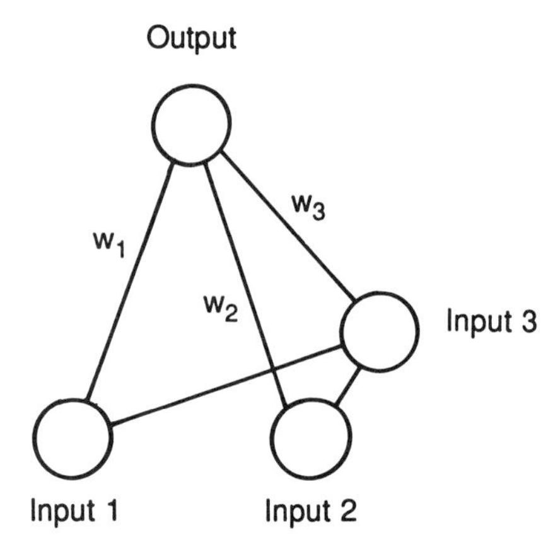

FIGURE 4. A multilayer network that converts the two-dimensional XOR problem into a three-dimensional linearly separable problem.

into three classes: *input units,* which receive the input patterns directly; *output units,* which have associated *teaching* or *target* inputs; and *hidden units,* which neither receive inputs directly nor are given direct feedback. This is the stock of units from which new features and new internal representations can be created. The problem is to know which new features are required to solve the problem at hand. In short, we must be able to learn intermediate layers. The question is, how? The original perceptron learning procedure does not apply to more than one layer. Minsky and Papert believed that no such general procedure could be found. To examine how such a procedure can be developed it is useful to consider the other major one-layer learning system of the 1950s and early 1960s, namely, the *least-mean-square (LMS)* learning procedure of Widrow and Hoff (1960).

Minimizing Mean Squared Error

The LMS procedure makes use of the delta rule for adjusting connection strengths; the perceptron convergence procedure is very similar, differing only in that linear threshold units are used instead of units with continuous-valued outputs. We use the term *LMS procedure* here to stress the fact that this family of learning rules may be viewed as minimizing a measure of the error in their performance.

The LMS procedure cannot be directly applied when the output units are linear threshold units (like the perceptron). It has been applied most often with purely linear output units. In this case the activation of an output unit, o_i, is simply given by $o_i = \sum_j w_{ij} i_j$. The error function, as indicated by the name least-mean-square, is the summed squared error. That is, the total error, E, is defined to be

$$E = \sum_p E_p = \sum_p \sum_i (t_{pi} - o_{pi})^2 \tag{1}$$

where the index p ranges over the set of input patterns, i ranges over the set of output units, and E_p represents the error on pattern p. The variable t_{pi} is the desired output, or *target,* for the ith output unit when the pth pattern has been presented, and o_{pi} is the actual output of the ith output unit when pattern p has been presented. The object is to find a set of weights that minimizes this function. It is useful to consider how the error varies as a function of any given weight in the system. Figure 5 illustrates the nature of this dependence. In the case of the simple single-layered linear system, we always get a smooth error function such as the one shown in the figure. The LMS procedure finds the values of all of the weights that minimize this function using a method called *gradient descent.* That is, after each pattern has been presented, the error on that pattern is computed and each weight is moved "down" the error gradient toward its minimum value for that pattern. Since we cannot map out the entire error function on each pattern presentation, we must find a simple procedure for determining, for each weight, how much to increase or decrease each weight. The idea of gradient descent is to make a change in the weight proportional to the

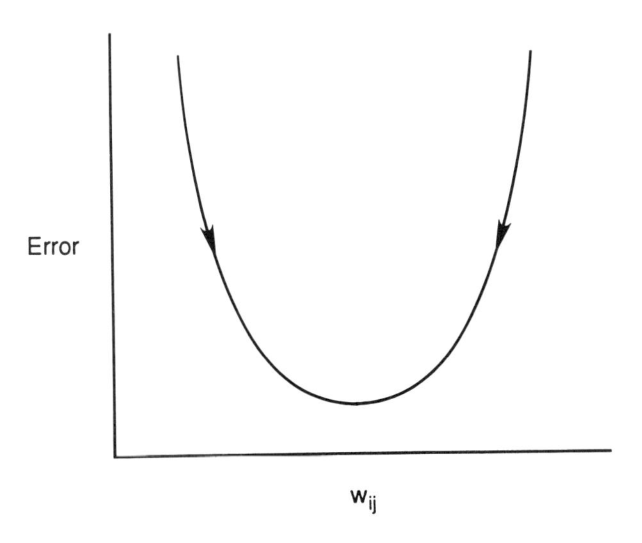

FIGURE 5. Typical curve showing the relationship between overall error and changes in a single weight in the network.

negative of the derivative of the error, as measured on the current pattern, with respect to each weight.[1] Thus the learning rule becomes

$$\Delta w_{ij} = -k \frac{\partial E_p}{\partial w_{ij}}$$

where k is the constant of proportionality. Interestingly, carrying out the derivative of the error measure in Equation 1 we get

$$\Delta w_{ij} = \epsilon \delta_{pi} i_{pj}$$

where $\epsilon = 2k$ and $\delta_{pi} = t_{pi} - o_{pi}$ is the difference between the target for unit i on pattern p and the actual output produced by the network. This is exactly the delta learning rule described in Equation 15 from Chapter 4. It should also be noted that this rule is essentially the same as that for the perceptron. In the perceptron the learning rate was 1 (i.e., we made unit changes in the weights) and the units were binary, but the rule itself is the same: the weights are changed proportionally to the difference between target and output times the input.

If we change each weight according to this rule, each weight is moved toward its own minimum and we think of the system as moving downhill in *weight-space* until it reaches its minimum error value. When all of the weights have reached their minimum points, the system has reached equilibrium. If the system is able to solve the problem entirely, the system will reach zero error and the weights will no longer be modified. On the other hand, if the network is unable to get the problem exactly right, it will find a set of weights that produces as small an error as possible.

In order to get a fuller understanding of this process it is useful to carefully consider the entire error space rather than a one-dimensional slice. In general this is very difficult to do because of the difficulty of depicting and visualizing high-dimensional spaces. However, we can usefully go from one to two dimensions by considering a network with exactly two weights. Consider, as an example, a linear network with two input units and one output unit with the task of finding a set of weights that comes as close as possible to performing the function OR. Assume the network has just two weights and no bias terms like the network in Figure 2A. We can then give some idea of the shape of the space by making a contour map of the error surface.

Figure 6 shows the contour map. In this case the space is shaped like a kind of oblong bowl. It is relatively flat on the bottom and rises sharply on the sides. Each equal error contour is elliptically shaped. The arrows

---

[1] It should be clear from Figure 5 why we want the negation of the derivative. If the weight is above the minimum value, the slope at that point is *positive* and we want to *decrease* the weight; thus when the slope is positive we add a negative amount to the weight. On the other hand, if the weight is too small, the error curve has a negative slope at that point, so we want to add a positive amount to the weight.

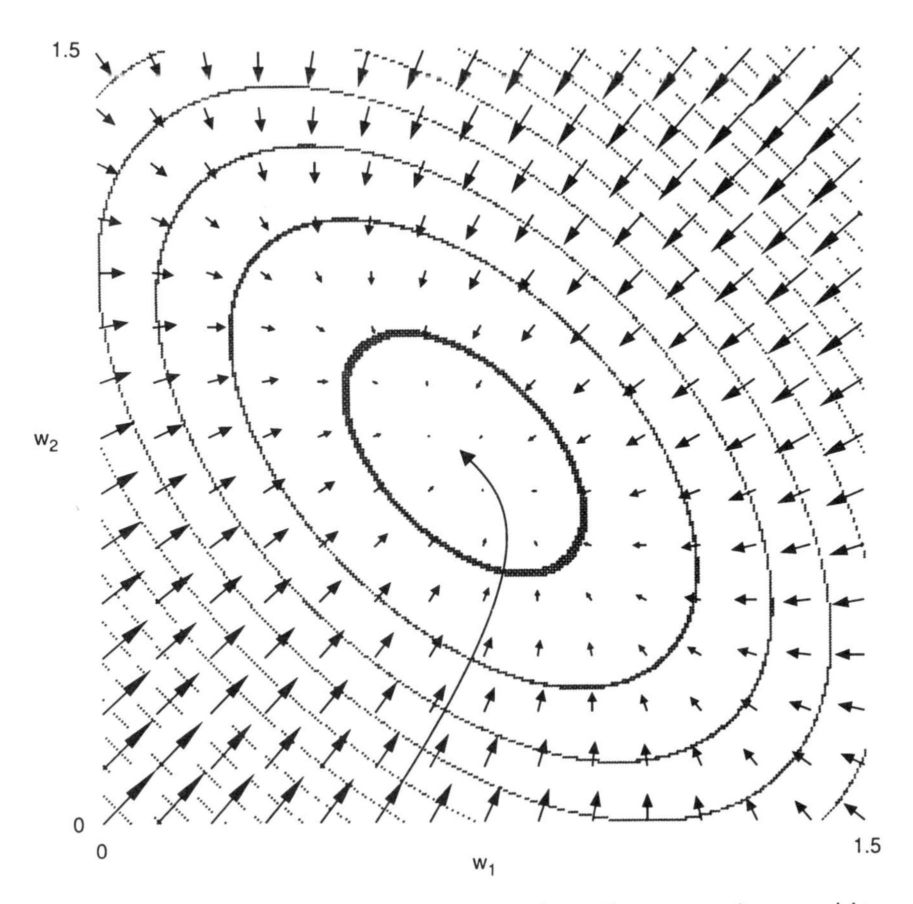

FIGURE 6. A contour map illustrating the error surface with respect to the two weights w_1 and w_2, for the OR problem in a linear network with two weights and no bias term. Note that the OR problem cannot be solved perfectly in a linear system. The minimum sum squared error over the four input-output pairs occurs when $w_1 = w_2 = 0.75$. (The input-output pairs are $00 \rightarrow 0$, $01 \rightarrow 1$, $10 \rightarrow 1$, and $11 \rightarrow 1$.)

around the ellipses represent the derivatives of the two weights at those points and thus represent the directions and magnitudes of weight changes at each point on the error surface. The changes are relatively large where the sides of the bowl are relatively steep and become smaller and smaller as we move into the central minimum. The long, curved arrow represents a typical trajectory in weight-space from a starting point far from the minimum down to the actual minimum in the space. The weights trace a curved trajectory following the arrows and crossing the contour lines at right angles.

The figure illustrates an important aspect of gradient descent learning. This is the fact that gradient descent involves making larger changes to parameters that will have the biggest effect on the measure being minimized. In this case, the LMS procedure makes changes to the weights proportional to the effect they will have on the summed squared error. The resulting total change to the weights is a vector that points in the direction in which the error drops most steeply.

The Back Propagation Rule

Although this simple linear pattern associator is a useful model for understanding the dynamics of gradient descent learning, it is not useful for solving problems such as the XOR problem mentioned above. As pointed out in *PDP:2*, linear systems cannot compute more in multiple layers than they can in a single layer. The basic idea of the back propagation method of learning is to combine a nonlinear perceptron-like system capable of making decisions with the objective error function of LMS and gradient descent. To do this, we must be able to readily compute the derivative of the error function with respect to *any weight in the network* and then change that weight according to the rule

$$\Delta w_{ij} = -k \frac{\partial E}{\partial w_{ij}}.$$

We will not bother with the mathematics here, since it is presented in detail in *PDP:8*. Suffice it to say, that with an appropriate choice of non-linear function we can perform the differentiation and derive the back propagation learning rule. The rule has exactly the same form as the learning rules described above, namely,[2]

$$\Delta w_{ij} = \epsilon \delta_{pi} a_{pj}.$$

The weight on each line should be changed by an amount proportional to the product of a term called δ, available to the unit receiving input along that line, times the activation, a, of the unit sending activation along that line.[3] The difference is in the exact determination of the δ term. Essentially, δ_{pi} represents the effect of a change in the net input to unit j on the output of unit i in pattern p. The determination of δ is a recursive process that starts with the output units. If a unit is an output unit, its δ is very similar to the one used in the standard delta rule. It is given by

$$\delta_{pi} = (t_{pi} - a_{pi}) f'_i(net_{pi})$$

[2] Note that the symbol η was used for the learning rate parameter in *PDP:8*. We use ϵ here for consistency with other chapters in this volume.

[3] In the networks we will be considering in this chapter, the output of a unit is equal to its activation. We use the symbol a to designate this variable. This symbol can be used for any unit, be it an input unit, an output unit, or a hidden unit.

where $net_{pi} = \sum_j w_{ij} a_{pj} + bias_i$ and $f'_i(net_{pi})$ is the derivative of the activation function with respect to a change in the net input to the unit. Note that $bias_i$ is a bias that has a similar function to the threshold, θ, in the perceptron.[4]

The δ term for hidden units for which there is no specified target is determined recursively in terms of the δ terms of the units to which it directly connects and the weights of those connections. That is,

$$\delta_{pi} = f'_i(net_{pi}) \sum_k \delta_{pk} w_{ki}$$

whenever the unit is not an output unit.

The application of the back propagation rule, then, involves two phases: During the first phase the input is presented and propagated forward through the network to compute the output value a_{pj} for each unit. This output is then compared with the target, resulting in a δ term for each output unit. The second phase involves a backward pass through the network (analogous to the initial forward pass) during which the δ term is computed for each unit in the network. This second, backward pass allows the recursive computation of δ as indicated above. Once these two phases are complete, we can compute, for each weight, the product of the δ term associated with the unit it projects to times the activation of the unit it projects from. Henceforth we will call this product the *weight error derivative* since it is proportional to (minus) the derivative of the error with respect to the weight. As will be discussed later, these weight error derivatives can then be used to compute actual weight changes on a pattern-by-pattern basis, or they may be accumulated over the ensemble of patterns.

The activation function. The derivation of the back propagation learning rule requires that the derivative of the activation function, $f'_i(net_i)$, exists. It is interesting to note that the linear threshold function, on which the perceptron is based, is discontinuous and hence will not suffice for back propagation. Similarly, since a linear system achieves no advantage from hidden units, a linear activation function will not suffice either. Thus, we need a continuous, nonlinear activation function. In most of our work on back propagation and in the program presented in this chapter, we have used the *logistic* activation function in which

$$a_{pi} = \frac{1}{1 + e^{-net_{pi}}}.$$

---

[4] Note that the values of the bias can be learned just like any other weights. We simply imagine that the bias is the weight from a unit that is always on.

In order to apply our learning rule, we need to know the derivative of this function with respect to its total input, net_{pi}. It is easy to show that this derivative is given by

$$\frac{da_{pi}}{dnet_{pi}} = a_{pi}(1 - a_{pi}).$$

Thus, for the logistic activation function, the error signal, δ_{pi}, for an output unit is given by

$$\delta_{pi} = (t_{pi} - a_{pi})a_{pi}(1 - a_{pi}),$$

and the error for an arbitrary hidden u_i is given by

$$\delta_{pi} = a_{pi}(1 - a_{pi})\sum_k \delta_{pk} w_{jk}.$$

It should be noted that the derivative, $a_{pi}(1 - a_{pi})$, reaches its maximum at $a_{pi} = 0.5$ and, since $0 \leqslant a_{pi} \leqslant 1$, approaches its minimum as a_{pi} approaches 0 or 1. Since the amount of change in a given weight is proportional to this derivative, weights will be changed most for those units that are near their midrange and, in some sense, not yet committed to being either on or off. This feature, we believe, contributes to the stability of the learning of the system.

Local minima. Like the simpler LMS learning paradigm, back propagation is a gradient descent procedure. Essentially, the system will follow the contour of the error surface—always moving downhill in the direction of steepest descent. This is no particular problem for the single-layer linear model. These systems always have bowl-shaped error surfaces. However, in multilayer networks there is the possibility of rather more complex surfaces with many minima. Some of the minima constitute solutions to the problems in which the system reaches an errorless state. All such minima are *global* minima. However, it is possible for some of the minima to be deeper than others. In this case, a gradient descent method may not find the *best* possible solution to the problem at hand. Part of the study of back propagation networks and learning involves a study of how frequently and under what conditions local minima occur. In problems with many hidden units, local minima seem quite rare. However with few hidden units, local minima can be more common. Figure 7 shows a very simple network in which we can demonstrate these phenomena. The network involves a single input unit, a single hidden unit, and a single output unit (a 1:1:1 network, for short). The problem is to copy the value of the input unit to the output unit. There are two basic ways in which the network can solve the problem. It can have positive biases on the hidden unit and on the output unit and large negative connections from the input unit to the hidden unit and from the hidden unit to the output unit, or it can have large negative biases on the two units and large positive weights from the input unit to the

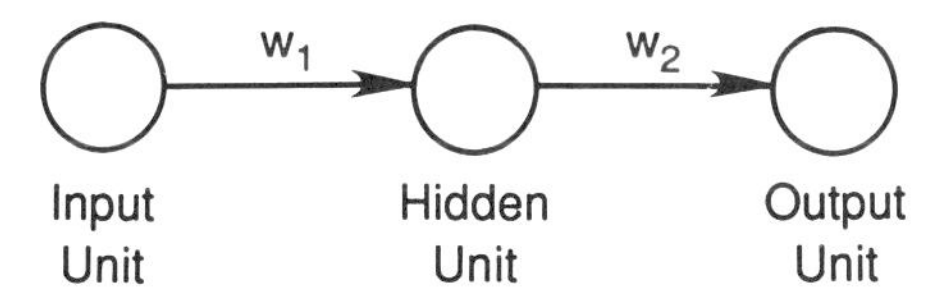

FIGURE 7. A 1:1:1 network, consisting of one input unit, one hidden unit, and one output unit.

hidden unit and from the hidden unit to the output unit. These solutions are illustrated in Table 3. In the first case, the solution works as follows: Imagine first that the input unit takes on a value of 0. In this case, there will be no activation from the input unit to the hidden unit, but the bias on the hidden unit will turn it on. Then the hidden unit has a *strong negative* connection to the output unit so it will be turned off, as required in this case. Now suppose that the input unit is set to 1. In this case, the strong inhibitory connection from the input to the hidden unit will turn the hidden unit off. Thus, no activation will flow from the hidden unit to the output unit. In this case, the positive bias on the output unit will turn it on and the problem will be solved. Now consider the second class of solutions. For this case, the connections among units are positive and the biases are negative. When the input unit is off, it cannot turn on the hidden unit. Since the hidden unit has a negative bias, it too will be off. The output unit, then, will not receive any input from the hidden unit and since its bias is negative, it too will turn off as required for zero input. Finally, if the input unit is turned on, the strong positive connection from the input unit to the hidden unit will turn on the hidden unit. This in turn will turn on the output unit as required. Thus we have, it appears, two symmetric solutions to the problem. Depending on the random starting state, the system will end up in one or the other of these *global* minima.

Interestingly, it is a simple matter to convert this problem to one with one local and one global minimum simply by setting the biases to 0 and not allowing them to change. In this case, the minima correspond to roughly the same two solutions as before. In one case, which is the global minimum as it turns out, both connections are large and negative. These minima are also illustrated in Table 3. Consider first what happens with

TABLE 3

WEIGHTS AND BIASES OF THE SOLUTIONS FOR A 1:1:1 NETWORK

Minima	w_1	w_2	$bias_1$	$bias_2$
Global	−8	−8	+4	+4
Global	+8	+8	−4	−4
Global	−8	−8	0	0
Local	+8	+0.73	0	0

both weights negative. When the input unit is turned off, the hidden unit receives no input. Since the bias is 0, the hidden unit has a net input of 0. A net input of 0 causes the hidden unit to take on a value of 0.5. The 0.5 input from the hidden unit, coupled with a large negative connection from the hidden unit to the output unit, is sufficient to turn off the output unit as required. On the other hand, when the input unit is turned on, it turns off the hidden unit. When the hidden unit is off, the output unit receives a net input of 0 and takes on a value of 0.5 rather than the desired value of 1.0. Thus there is an error of 0.5 and a squared error of 0.25. This, it turns out, is the best the system can do with zero biases. Now consider what happens if both connections are positive. When the input unit is off, the hidden unit takes on a value of 0.5. Since the output is intended to be 0 in this case, there is pressure for the weight from the hidden unit to the output unit to be small. On the other hand, when the input unit is on, it turns on the hidden unit. Since the output unit is to be on in this case, there is pressure for the weight to be large so it can turn on the output unit. In fact, these two pressures balance off and the system finds a compromise value of about 0.73. This compromise yields a summed squared error of about 0.45—a local minima.

Usually, it is difficult to see why a network has been caught in a local minimum. However, in this very simple case, we have only two weights and can produce a contour map for the error space. The map is shown in Figure 8. It is perhaps difficult to visualize, but the map roughly shows a saddle shape. It is high on the upper left and lower right and slopes down toward the center. It then slopes off on each side toward the two minima. If the initial values of the weights begin below the antidiagonal (that is, below the line $w_1 + w_2 = 0$), the system will follow the contours down and to the left into the minimum in which both weights are negative. If, however, the system begins above the antidiagonal, the system will follow the slope into the upper right quadrant in which both weights are positive. Eventually, the system moves into a gently sloping valley in which the weight from the hidden unit to the output unit is almost constant at about 0.73 and the weight from the input unit to the hidden unit is slowly increasing. It is slowly being sucked into a local minimum. The directed arrows superimposed on the map illustrate the lines of force and illustrate these dynamics. The long arrows represent two trajectories through weight-space for two different starting points.

It is rare that we can create such a simple illustration of the dynamics of weight-spaces and how local minima come about. However, it is likely that many of our spaces contain these kinds of saddle-shaped error surfaces. Sometimes, as when the biases are free to move, there is a global minimum on either side of the saddle point. In this case, it doesn't matter which way you move off. At other times, such as in Figure 8, the two sides are of different depths. There is no way the system can sense the depth of a minimum from the edge, and once it has slipped in there is no way out. Importantly, however, we find that high-dimensional spaces (with many

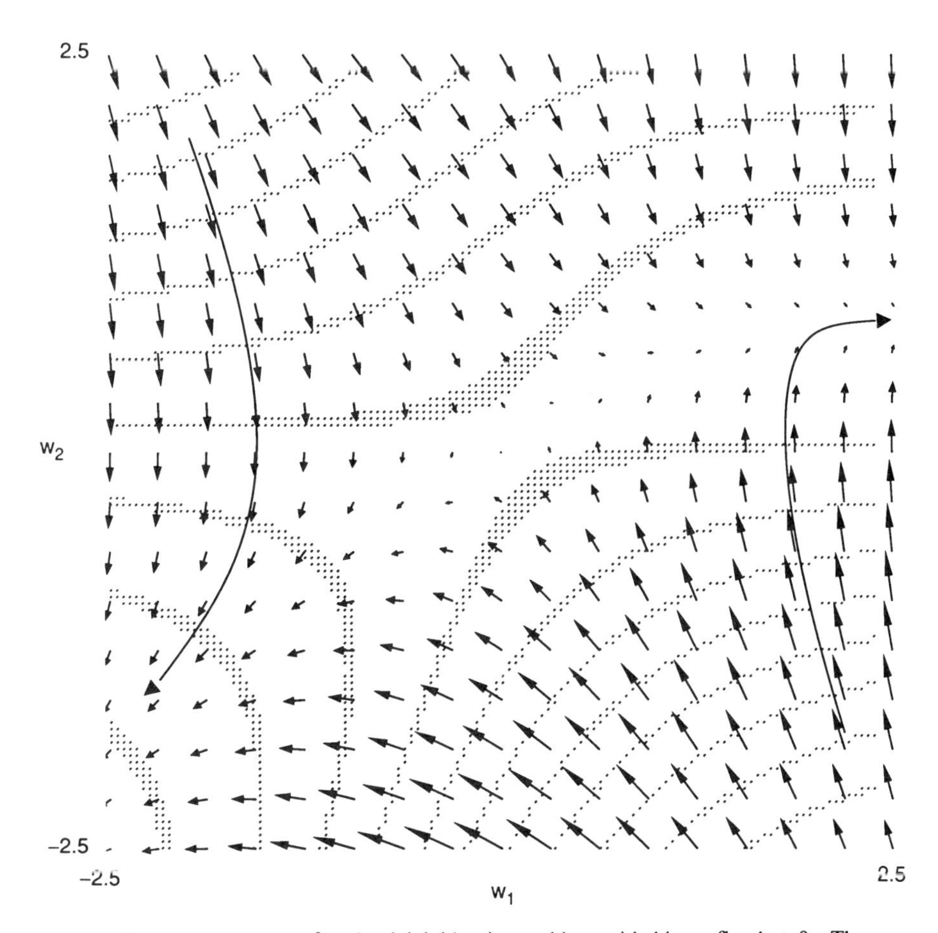

FIGURE 8. A contour map for the 1:1:1 identity problem with biases fixed at 0. The map show a local minimum in the positive quadrant and a global minimum in the lower left-hand negative quadrant. Overall the error surface is saddle-shaped. See the text for further explanation.

weights) have relatively few local minima. It seems that the system can always, as it were, slip along another dimension to find a path out of most local minima.

Momentum. Our learning procedure requires only that the change in weight be proportional to the weight error derivative. True gradient descent requires that infinitesimal steps be taken. The constant of proportionality, ϵ, is the learning rate in our procedure. The larger this constant, the larger the changes in the weights. For practical purposes we choose a learning rate that is as large as possible without leading to oscillation. This offers the most rapid learning. One way to increase the learning rate without leading to oscillation is to modify the back propagation learning rule to include a *momentum* term. This can be accomplished by the following rule:

$$\Delta w_{ij}(n+1) = \epsilon(\delta_{pi}a_{pj}) + \alpha\Delta w_{ij}(n)$$

where the subscript n indexes the presentation number and α is a constant that determines the effect of past weight changes on the current direction of movement in weight space. This provides a kind of momentum in weight-space that effectively filters out high-frequency variations of the error surface in the weight-space. This is useful in spaces containing long ravines that are characterized by sharp curvature across the ravine and a gently sloping floor. The sharp curvature tends to cause divergent oscillations across the ravine. To prevent these it is necessary to take very small steps, but this causes very slow progress along the ravine. The momentum filters out the high curvature and thus allows the effective weight steps to be bigger. In most of the simulations reported in *PDP:8*, α was about 0.9. Our experience has been that we get the same solutions by setting $\alpha = 0$ and reducing the size of ϵ, but the system learns much faster overall with larger values of α and ϵ.

Symmetry breaking. Our learning procedure has one more problem that can be readily overcome and this is the problem of symmetry breaking. If all weights start out with equal values and if the solution requires that unequal weights be developed, the system can never learn. This is because error is propagated back through the weights in proportion to the values of the weights. This means that all hidden units connected directly to the output units will get identical error signals, and, since the weight changes depend on the error signals, the weights from those units to the output units must always be the same. The system is starting out at a kind of unstable equilibrium point that keeps the weights equal, but it is higher than some neighboring points on the error surface, and once it moves away to one of these points, it will never return. We counteract this problem by starting the system with small random weights. Under these conditions symmetry problems of this kind do not arise. This can be seen in Figure 8. If the system starts at exactly (0,0), there is no pressure for it to move at all and the system will not learn, but if it starts anywhere off of the antidiagonal, it will eventually end up in one minimum or the other.

Learning by pattern or by epoch. The derivation of the back propagation paradigm supposes that we are taking the derivative of the error function summed over all patterns. In this case, we might imagine that we would present all patterns and then sum the derivatives before changing the weights. Instead, we can compute the derivatives on each pattern and make the changes to the weights after each pattern rather than after each epoch. If the learning rate is small, there is little difference between the two procedures, and the version in which weights are changed after each pattern seems more satisfying—since there might be a very large set of patterns. The **bp** program introduced in the next section allows both of these options.

IMPLEMENTATION

The **bp** program implements the back propagation process just described. The program makes use of a network specification file, which indicates the architecture of the network. Networks are assumed to be feedforward only, with no recurrence, although limited recurrence can be simulated by linking weights to each other or by copying activations of output units back into the input units. These variations are described at the end of this chapter.

The network specifications indicate how many total units are in the network and, of these, how many are input units and how many are output units. The units in the network specification are assumed to be ordered in such a way that no unit ever receives input from a unit occurring later in the ordering. Thus, the first listed units are the input units, the next listed are the hidden units, and the last are the output units. Among the hidden units, if there are multiple layers, the units in the layer closest to the input are listed first, and so on.

Weights may constrained to be positive, negative, or linked. When a weight is "linked" that means that it is forced to have the same value as all of the other weights that it is linked to; there may be any number of "link groups" of weights. These constraints are imposed each time the weights are incremented during processing. Two other constraints are imposed only when weights are initialized; these constraints are either a fixed (floating-point) value to which the weight is initialized or a random value. For weights that are random, if they are constrained to be positive, they are initialized to a value between 0 and the value of a parameter called *wrange*; if the weights are constrained to be negative, the initialization value is between $-wrange$ and 0; otherwise, the initialization value is between $wrange/2$ and $-wrange/2$. Weights that are constrained to a fixed value are initialized to that value. Bias terms are treated like weights and may be subject to all the same kinds of constraints.

The program also allows the user to set individually for each weight and bias term whether the weight or bias will be modifiable. The weight modifiability information is recorded in a matrix called *epsilon*; there is also a corresponding vector called *bepsilon* that records the modifiability of the bias terms of each unit. When the network is initialized, *epsilon* and *bepsilon* are set to *lrate* (the value of the global learning rate parameter) or 0, depending on whether the weight is modifiable or not. If *lrate* is changed, all nonzero *epsilon* and *bepsilon* terms are set to the new value of *lrate*. (The conventions for setting up a network specification file are described in Appendix C.)

The **bp** program also makes use of a list of pattern pairs, each pair consisting of a name, an input pattern, and a target pattern.

Processing of a single pattern occurs as follows: A pattern pair is chosen, and the pattern of activation specified by the input pattern is clamped on the input units; that is, their activations are set to 1 or 0 based on the values found in the input pattern.

Next, activations are computed. For each noninput unit, the net input to the unit is computed and then the activation of the unit is set. This occurs in the order that the units are specified in the network specification, so that by the time each unit is encountered, the activations of all of the units that feed into it have already been set. The routine that performs this computation is

```
compute_output() {

  for (i = ninputs; i < nunits; i++) {
    netinput[i] = bias[i];
    for (j=first_weight_to[i]; j<last_weight_to[i]; j++) {
      netinput[i] += activation[j]*weight[i][j];
    }
    activation[i] = logistic(netinput[i]);
  }
}
```

In this code, note that *ninputs*, which is the number of input units, is also the index of the first hidden unit. The arrays *first_weight_to* and *last_weight_to* indicate which unit is the first and which is the last to project to each unit.[5]

Next, *error* and *delta* terms are computed. The *error* for a unit is equivalent to the partial derivative of the error with respect to a change in the *activation* of the unit. The *delta* for the unit is the partial derivative of the error with respect to a change in the *net input* to the unit.

First, the *delta* and *error* terms for all units are set to 0. Then, *error* terms are calculated for each output unit. For these units, *error* is the difference between the target and the obtained activation of the unit.

After the error has been computed for each output unit, we get to the "heart" of back propagation: the recursive computation of *error* and *delta* terms for hidden units. The program iterates backward over the units, starting with the last output unit. The first thing it does in each pass through the loop is set *delta* for the current unit, which is equal to the *error* for the unit times the derivative of the activation function (i.e., the activation of the unit times one minus its activation). Then, once it has *delta* for the current unit, the program passes this back to all units that have

[5] Note that for efficiency and other reasons, the weight indexes are not actually implemented in the form shown here. A description of the actual treatment of weight arrays is given in Appendix F.

connections coming into the current unit; this is the actual back propagation process. By the time a particular unit becomes the current unit, all of the units that it projects to will have already been processed, and all of its error will have been accumulated, so it is ready to have its *delta* computed. The code for this is as follows:

```
compute_error() {

  for (i = ninputs; i < nunits; i++) {
    error[i] = 0.0;
  }

  for (i = nunits-noutputs, t=0; i<nunits; t++, i++) {
    error[i] = target[t] - activation[i];
  }

  for (i = nunits - 1; i >= ninputs; i--) {
    delta[i] = error[i]*activation[i]*(1.0-activation[i]);
    for (j=first_weight_to[i]; j<last_weight_to[i]; j++)
      error[j] += delta[i] * weight[i][j];
  }
}
```

After computing error, if the *lflag* is nonzero, the weight error derivatives are then computed from the *delta*s and *activation*s. The error derivatives for the *bias* terms are also computed. (Recall that the *bias* terms are equivalent to weights to a unit from a unit whose activation is always 1.0.) These computations occur in the following routine:

```
compute_wed() {

  for (i = ninputs; i < nunits; i++) {
    for (j=first_weight_to[i]; j<last_weight_to[i]; j++) {
      wed[i][j] += delta[i] * activation[j];
    }
    bed[i] += delta[i];
  }
}
```

Note that this routine adds the weight error derivatives occasioned by the present pattern into an array where they can potentially be accumulated over patterns.

Weight error derivatives actually lead to changes in the weights either after processing each pattern or after each entire epoch of processing. In either case, the computation that is actually performed needs to be clearly understood. For each weight, a delta weight is first calculated. The delta weight is equal to the accumulated weight error derivative plus a fraction of the previous delta weight, where the size of the fraction is determined by the parameter *momentum*. Then, this delta weight is added into the weight, so that the weight's new value is equal to its old value plus the delta weight. Again, the same computation is performed for all of the *bias* terms. The following routine performs these computations:

```
change_weights() {

  sum_linked_weds();

  for (i = ninputs; i < nunits; i++) {
    for (j=first_weight_to[i]; j<last_weight_to[i]; j++) {
      dweight[i][j] = epsilon[i][j]*wed[i][j] +
                         momentum*dweight[i][j];
      weight[i][j] += dweight[i][j];
      wed[i][j] = 0.0;
    }
    dbias[i] = bepsilon[i]*bed[i] + momentum*dbias[i];
    bias[i] += dbias[i];
    bed[i] = 0.0;
  }

  constrain_neg_pos();
}
```

Note that before the weights are actually changed, the *sum_linked_weds* routine is called. This routine adds together all the weight error derivative terms associated with all the weights that are linked together in the same link group. The idea of linking weights is to assure that all weights that are linked together always have the same value since conceptually they are thought of as being a single weight. Also, after the weights are changed, the *constrain_neg_pos* routine is called to make sure that the values assigned to the weights conform to the positive or negative constraints that have been imposed on them in the network specification file. Weights that are constrained to be positive are reset to 0 by *constrain_neg_pos* if *change_weights* trys to put them below 0, and weights that are constrained to be negative are reset to 0 if *change_weights* tries to put them above 0.

We have just described the processing activity that takes place for each input-target pair in each learning *trial*. Generally, learning is accomplished through a sequence of *epochs*, in which all pattern pairs are presented for one trial each during each epoch. The *change_weights* routine may be called once per pattern or only once per epoch. The presentation is either in sequential or permuted order. It is also possible to test the processing of patterns, either individually or by sequentially cycling through the whole list, with learning turned off. In this case, *compute_output* and *compute_error* are called, but *compute_wed* and *change_weights* are not called.

Whether or not learning is occurring, the program also computes summary statistics after processing each pattern. First it computes the pattern sum of squares (*pss*), equal to the squared error terms summed over all of the output units. Then it adds the *pss* to the total sum of squares (*tss*), which is just the cumulative sum of the *pss* for all patterns thus far processed within the current epoch.

Learning is carried out by the *strain* and *ptrain* commands. The first carries out training in sequential order, the second in permuted order. Training goes on for *nepochs* or until the value of *tss* becomes less than the value of a control parameter called *ecrit* for "error criterion." (Note that *strain* and *ptrain* do not check the *tss* until the end of each epoch, so they will always run through at least one epoch before returning.)

Modes and Measures

The basic back propagation procedure can be operated in either of two *learning grain* modes: *pattern* or *epoch*. If the learning grain is set to *pattern*, the weight error derivatives are computed for each pattern, and then *change_weights* is called, updating the delta weights and adding them to the weights before the next pattern is processed. If learning grain is set to *epoch*, the weight error derivatives are still computed for each pattern, but they are accumulated until the end of the epoch. In this case, then, delta weights are computed and added into the actual weight array only once per epoch.

In the **bp** program the principle measure of performance is the pattern sum of squares (*pss*) and the total sum of squares (*tss*). The user may optionally also compute an additional measure, the vector correlation of the present weight error derivatives with the previous weight error derivatives. The set of weight error derivatives can be thought of as a vector pointing in the steepest direction downhill in weight space; that is, it points down the error gradient in weight space. Thus, the vector correlation of these derivatives across successive epochs indicates whether the gradient is staying relatively stable or shifting from epoch to epoch. For example, a negative value of this correlation measure (called *gcor* for *gradient correlation*) indicates that the gradient is changing in direction. Since the *gcor* can be thought of as following the gradient, the mode switch for turning on this computation is called *follow*.

The only other mode available in the **bp** program is *cascade* mode. This mode allows activation to build up gradually rather than being computed in a single step as is usually the case in **bp**. A discussion of the implementation and use of this mode is provided later in the "Extensions" section.

RUNNING THE PROGRAM

The **bp** program is used much like earlier programs in this series, particularly **pa**. Like the other programs, it has a flexible architecture that must be specified using a *.net* file. The conventions used in these files are described in Appendix C. The program also makes use of a *.pat* file, in which the pairs of patterns to be used in training and testing the network are listed.

When networks are initialized, the weights are generally assigned according to a pseudo-random number generator. As in **pa** and **iac**, the *reset* command allows the user to repeat a simulation run with the same initial configuration used just before. (Another procedure for repeating the previous run is described in Ex. 5.1.) The *newstart* command generates a new random seed and seeds the random number generator with it.

Commands

All of the commands in **bp** will be familiar from previous programs. The only ones that are notably changed in usage from **pa** are the *test* and *get/ patterns* commands. In **bp**, *test* simply prompts for the name or number of a pattern from the pattern list and does not allow the option of specifying an arbitrary input pattern; *get/ patterns* works as before except that negative entries in the *.pat* file are treated specially in **bp**; see the *env/ ipattern* and *env/ tpattern* entries in the list of variables below.

Variables

There are several new variables and a few minor changes to the meaning of some familiar variables. These new and changed variables are described in the following list. As in previous programs, all of the variables in **bp** are accessed via the *set/* and *exam/* commands.

ncycles
: The number of processing cycles run when the *cycle* routine is called. This program control variable is used in *cascade* mode, described later in this chapter.

stepsize
: Size of the processing step taken before updating the screen and prompting for *return* in single step mode. Possible values are *cycle*, *ncycles*, *pattern*, *epoch*, and *nepochs*. When *cascade* mode is not on and if the value of *stepsize* is *cycle* or *ncycles*, the screen is updated after the forward processing sweep, then again after the backward processing sweep and any weight adjustments. When *cascade* mode is on and if the *stepsize* is *cycle*, updating occurs after each processing cycle; if *stepsize* is *ncycles*, updating occurs after *ncycles* of processing.

config/ bepsilon

For each bias term, there is an associated *bepsilon*, or modifiability parameter, just as for each weight. These are just like *epsilons* (see below), except that the user must only indicate one index since there is just one per unit.

config/ epsilon

For each weight, there is an associated *epsilon*, or modifiability parameter. Generally, *epsilon[i][j]* (to unit *i* from unit *j*) is equal to either *lrate* or 0.0, according to whether *weight[i][j]* is modifiable, as specified in the *.net* file. When *lrate* is changed, all the nonzero *epsilons* are set to *lrate*. However, the user may then independently adjust *epsilon* on any particular weight to any value desired. The user must specify both a receiver and a sender to indicate which *epsilon* to examine or change.

env/ ipattern

Input patterns are specified as a sequence of floating-point numbers, as in earlier programs. The entries "+", ".", and "−" are allowed shorthand for +1.0, 0.0, and −1.0, respectively. Legal values are between 0 and 1, inclusive. If an element of an input pattern has a negative value, the activation of the corresponding input unit is set to the activation of the unit whose index is the absolute value of the negative input pattern element. If the parameter *mu* is nonzero, the activation value of the input unit is incremented by *mu* times its previous value. For example, if element 3 of a particular input pattern is −12, the activation of input unit 3 is set equal to the activation calculated for input unit 12 in processing the previous pattern, plus *mu* times the previous activation of input unit 3.

env/ tpattern

Target patterns are specified as a sequence of floating-point numbers as in **pa**. The entries "+", "−", and "." are interpreted as in the *ipattern*. Legal target values are between 0 and 1, inclusive. If an element of a target pattern has a negative value, then the program acts as though no target was specified for the corresponding output unit; the error for that unit is therefore set to 0.0.

mode/ cascade

By default, the forward pass of processing occurs in a single sweep in **bp**. If this mode variable is set to 1, however, net inputs accumulate gradually over several processing cycles. In this mode the *ncycles* variable determines the number of processing cycles run per pattern tested, and the *cycle* command, following the *test* command, allows the user to continue cycling.

mode/ follow

When this mode switch is set to 1, the **bp** program computes the *gcor* measure, which is the correlation of the gradient in weight space between successive calls to *change_weights*.

mode/ lgrain

Refers to the grain of learning or weight adjustment. By default in the program it is set to *pattern*, which means that weights are adjusted after processing each pattern pair. The *lgrain* variable may also be set to *epoch*, in which case weight changes are accumulated over all patterns presented within an epoch, and then the weights are actually changed only at the end of the epoch.

param/ crate

The cascade rate parameter. Default value is 0.05. Determines the rate of build-up of the net input to each unit on each processing cycle.

param/ lrate

The learning rate parameter. Its default value is 0.05. When *lrate* is reset, all *epsilon*s and *bepsilon*s that are nonzero are reset to the new value of *lrate*.

param/ momentum

The value of the momentum parameter (called *alpha* in *PDP:8*); it has a default value of 0.9. Note that when *lgrain* is set to *pattern*, momentum is building up over patterns; when *lgrain* is *epoch*, momentum is building up over epochs.

param/ mu

This parameter applies to the extension of **bp** to sequential networks. It specifies the extent to which the previous activation of a unit is averaged together with input it receives from other units.

param/ tmax

The target activation actually used when the value given in the target pattern is 1. Defaults to 1.0. The target activation used when the target string contains a 0 is $(1 - tmax)$, which is 0.0 when *tmax* is 1.0.

param/ wrange

Range of variability for random weights. If constrained to be positive, the random weights range from 0 to $+wrange$. If constrained to be negative, they range from $-wrange$ to 0. If unconstrained, they range from $-wrange/2$ to $+wrange/2$.

state/ activation

Vector of activation values for units. The ith element corresponds to the activation of the ith unit, as computed during the most recent processing cycle.

state/ bed

Vector of bias error derivative terms.

state/ dbias

Vector of delta bias terms, comparable to delta weights, see below.

state/ delta

Vector of delta terms, or partial derivatives of the error with respect to the net inputs.

state/ dweight
Matrix of delta weights (Δw_{ij} from the equations). This matrix contains the increment last added to the weights, and is used when momentum is nonzero in setting the value of the next increment to add.

state/ error
Vector of error terms, or the partial derivative of the error with respect to a change in the activation of each unit.

state/ gcor
Measures the correlation of the direction of the gradient in weight-space between successive calls to *change_weights*. This is computed only if *follow* mode is on.

state/ netinput
Net input vector.

state/ target
Vector of targets. Note that the ith element corresponds to target value for ith output unit.

state/ wed
Matrix of weight error derivative terms.

OVERVIEW OF EXERCISES

We present three exercises using the basic back propagation procedure. The first one takes you through the XOR problem and is intended to allow you to test and consolidate your basic understanding of the back propagation procedure and the gradient descent process it implements. The second exercise suggests minor variations of the basic back propagation procedure, such as whether weights are changed pattern by pattern or epoch by epoch, and also proposes various parameters that may be explored. The third exercise suggests other possible problems that you might want to explore. Several further exercises are described later in the chapter, in the section on extensions of the back propagation procedure. In Appendix E we provide answers for the questions in Ex. 1.

Ex. 5.1. The XOR Problem

The XOR problem is described at length in *PDP:8*. Here we will be considering one of the two network architectures considered there. This architecture is shown in Figure 9. In this network configuration there are two input units, one for each "bit" in the input pattern; two hidden units; and one output unit. The input units project to the hidden units, and the

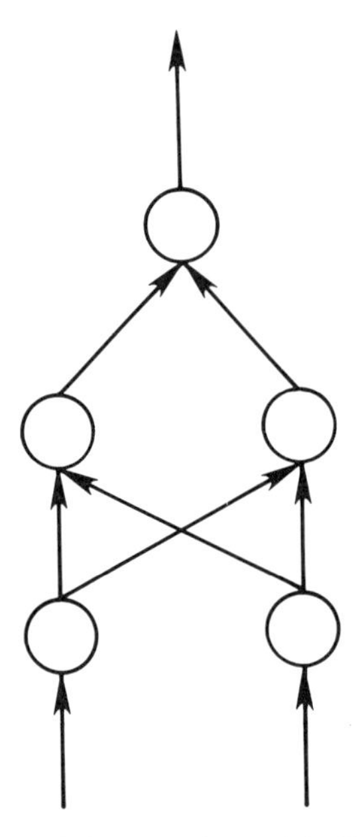

FIGURE 9. Architecture of the XOR network used in Exs. 5.1 and 5.2. (From *PDP:8*, p. 332.)

hidden units project to the output unit; there are no direct connections from the input units to the output units.

All of the relevant files for doing this exercise are contained in the *bp* directory; they are called *xor.tem, xor.str, xor.net*, *xor.pat*, and *xor.wts*. These contain the template, start-up, network, and pattern specifications needed for running the XOR problem.

Once you have your own directory set up with copies of the relevant files, you can start the program. To do so, type

bp xor.tem xor.str

If you wish, you may use a different template file, called *xor2.tem*, instead of *xor.tem*. We describe the case where *xor.tem* is used because the screen layout used in this file is easier to modify for other problems. The layout used in *xor2.tem* should be self-explanatory.

The *xor.str* file is as follows:

```
get network xor.net
get patterns xor.pat
set nepochs 30
set ecrit 0.04
set dlevel 3
```

```
set slevel 1
set lflag 1
set mode lgrain epoch
set mode follow 1
set param lrate 0.5
get weights xor.wts
tall
```

This file instructs the program to set up the network as specified in the *xor.net* file and to read the patterns as specified in the *xor.pat* file; it also initializes various variables. Then it reads in an initial set of weights to use for this exercise. Finally, *tall* is called, so that the program processes each pattern. The display that finally appears shows the state of the network at the end of this initial test of all of the patterns. It is shown in Figure 10.

In this figure, below the prompt and the top-level menu and to the left, the current epoch number and the total sum of squares (*tss*) resulting from testing all four patterns are displayed. Also displayed is the *gcor* measure, which is 0.0 at this point since no weight error derivatives have been computed. The next line contains the current pattern number and the sum of squares associated with this pattern. To the right of these entries is the set of input and target patterns for XOR. Below all these entries is a horizontal vector indicating the activations of all of the "sender units." These are units that send their activations forward to other units in the network. The first two are the two input units, and the next two are the two hidden units.

Below this row vector of sender activations is the matrix of weights. The weight in a particular column and row represents the strength of the connection from a particular sender unit indexed by the column to the particular receiver indexed by the row. Note that only the weights that actually

```
bp:
disp/  exam/  get/  save/  set/  clear  cycle  do  log  newstart  ptrain  quit
reset  run  strain  tall  test

                                          pname  ipatterns   tpatterns
epoch        0    tss       1.0507         p00      0 0          0
                  gcor      0.0000         p01      0 1          1
cpname     p11    pss       0.3839         p10      1 0          1
                                           p11      1 1          0

sender acts:          100 100   64   40     bia net act tar del

weights:               43  44                27  60  64        9
                        3   3                40  40  40        2
                               27    8       27  48  61    0 146
```

FIGURE 10. The display produced by **bp**, initialized for XOR.

exist in the network are displayed—these are the weights from the two input units to the two hidden units (these are the four numbers below the activations of the two input units) and the weights from the two hidden units to the single output unit (these are the numbers below the activations of the two sending units).

To the right of the weights is a column vector indicating the values of the bias terms for the *receiver* units—that is, all the units that receive input from other units. In this case, the receivers are the two hidden units and the output unit.

To the right of the biases is a column for the net input to each receiving unit. There is also a column for the activations of each of these receiver units. (Note that the hidden units' activations appear twice, once in the row of senders and once in this column of receivers.) The next column contains the target vector, which in this case has only one element since there is only one output unit. Finally, the last column contains the delta values for the hidden and output units.

Note that all activation, weight, and bias values are given in hundredths, so that, for example, 43 means 0.43 and reverse-video 3 means −0.03. For deltas, values are given in *thousandths* so that reverse-video 9 means −0.009.

The display shows what happened when the last pattern pair in the file *xor.pat* was processed. This pattern pair consists of the input pattern (1 1) and the target pattern (0). This input pattern was clamped on the two input units. This is why they both have activation values of 1.0, shown as 100 in the first two entries of the sender activation vector. With these activations of the input units, coupled with the weights from these units to the hidden units, and with the values of the bias terms, the net inputs to the hidden units were set to 0.60 and −0.40, as indicated in the *net* column. Plugging these values into the logistic function, the activation values of 0.64 and 0.40 were obtained for these units. These values are recorded both in the sender activation vector and in the receiver activation vector (labeled *act*, next to the net input vector). Given these activations for the hidden units, coupled with the weights from the hidden units to the output unit and the bias on the output unit, the net input to the output unit is 0.48, as indicated at the bottom of the *net* column. This leads to an activation of 0.61, as shown in the last entry of the *act* column. Since the target is 0.0, as indicated in the target column, the *error*, or (*target* − *activation*) is −0.61; this error, times the derivative of the activation function (that is, *activation* (1 − *activation*)) results in a delta value of −0.146, as indicated in the last entry of the final column. The delta values of the hidden units are determined by back propagating this delta term to the hidden units, as specified by the *compute_error* subroutine.

Q.5.1.1. Show the calculations of the values of *delta* for each of the two hidden units, using the activations and weights as given in this initial screen display. Explain why these values are so small.

At this point, you will notice that the total sum of squares before any learning has occurred is 1.0507. Run another *tall* to understand more about what is happening.

Q.5.1.2. Report the output the network produces for each input pattern and explain why the values are all so similar, referring to the strengths of the weights, the logistic function, and the effects of passing activation forward through the hidden units before it reaches the output units.

Now you are ready to begin learning. Use the *strain* command. This will run 30 epochs of training because the *nepochs* variable is set to 30. If you set *single* to 1, you can follow the *tss* and *gcor* measures as they change from epoch to epoch. You may find in the course of running this exercise that you need to go back and start again. To do this, you should use the *reset* command, followed by

get weights xor.wts

The *xor.wts* file contains the initial weights used for this exercise. This method of reinitializing guarantees that all users will get the same starting weights, independent of possible differences in random number generators from system to system.

Q.5.1.3. The total sum of squares is smaller at the end of 30 epochs, but is only a little smaller. Describe what has happened to the weights and biases and the resulting effects on the activation of the output units. Note the small sizes of the deltas for the hidden units and explain. Do you expect learning to proceed quickly or slowly from this point? Why?

Run another 90 epochs of training (for a total of 120) and see if your predictions are confirmed. As you go along, keep a record of the *tss* at each 30-epoch milestone. (The initial value is given in Figure 10, in case you did not record this previously.) You might find it interesting to observe the results of processing each pattern rather than just the last pattern in the four-pattern set. To do this, you can set the *stepsize* variable to *pattern* rather than the default *epoch*.

At the end of another 60 epochs (total: 180), some of the weights in the network have begun to build up. At this point, one of the hidden units is providing a fairly sensitive index of the number of input units that are on. The other is very unresponsive.

Q.5.1.4. Explain why the more responsive hidden unit will continue to change its incoming weights more rapidly than the other unit over the next few epochs.

Run another 30 epochs. At this point, after a total of 210 epochs, one of the hidden units is now acting rather like an OR unit: its output is about the same for all input patterns in which one or more input units is on.

Q.5.1.5. Explain this OR unit in terms of its incoming weights and bias term. What is the other unit doing at this point?

Now run another 30 epochs. During these epochs, you will see that the second hidden unit becomes more differentiated in its response.

Q.5.1.6. Describe what the second hidden unit is doing at this point, and explain why it is leading the network to activate the output unit most strongly when only one of the two input units is on.

Run another 30 epochs. Here you will see the *tss* drop very quickly.

Q.5.1.7. Explain the rapid drop in the *tss*, referring to the forces operating on the second hidden unit and the change in its behavior. Note that the size of the *delta* for this hidden unit at the end of 270 epochs is about as large in absolute magnitude as the size of the *delta* for the output unit. Explain.

Run the *strain* command one more time. Before the end of the 30 epochs, the value of *tss* drops below *ecrit*, and so *strain* returns. The XOR problem is solved at this point.

Q.5.1.8. Summarize the course of learning, and compare the final state of the weights with their initial state. Can you give an approximate intuitive account of what has happened? What suggestions might you make for improving performance based on this analysis?

Ex. 5.2. Variations With XOR

There are several further studies one can do with XOR. You can study the effects of varying:

- One of the parameters of the model (*lrate, wrange, momentum*).
- The learning grain (*lgrain* [*epoch* vs. *pattern*]).
- The training regime (permuted vs. sequential test; this makes a difference only when *lgrain* is equal to *pattern*).

- The starting configuration (affected by *wrange*, but also by the particular random values of the weights).

- The number of hidden units.

- Whether particular weights are constrained to be positive, negative, or linked to each other.

The possibilities are almost endless, and all of them have effects. For this exercise, pick one of these possible dimensions of variation, and run at least three more runs, comparing the results to those you obtained in Ex. 5.1.

Q.5.2.1. Describe what you have chosen to vary, how you chose to vary it, and what results you obtained in terms of the rate of learning, the evolution of the weights, and the eventual solution achieved.

Where relevant, it is useful to try several different values along the dimension you have chosen to vary, performing several runs with different initial starting weights in each instance and using the *newstart* command to get a new value for the random seed before each run. You might want to try the same starting configuration with different values on the dimension you are varying. This can be done using *reset* instead of *newstart*. For example, suppose you are examining the effects of varying *lrate*. You can issue the *newstart* command to get a random weight set to use with one value of *lrate*, then use *reset* to try the same starting configuration again, after using the *set/ param/ lrate* command to adjust the value of *lrate*.

If you want to constrain weights to be positive, negative, or linked, you need only change the *.net* file a little, replacing *r*'s with *p*'s or *n*'s for weights that you want to force to be positive or negative. It is best to make a new *.net* file, with a new name, and a new *.str* file that includes the *get/ network* command to read in this new *.net* file.

If you choose to vary the number of hidden units, you will have to modify the *.net* and *.tem* files. Here we give very briefly a checklist of what needs to be done to accomplish this. For the network file, you must increase *nunits* in the definitions. You must also alter the network part of the file and the biases part of the file to specify the values of the connections and bias terms involving the added units. Since *xor.net* uses the % convention, you will have to understand how it works, as described in Appendix C. You have to change the %-lines to reflect the new hidden units.

For the template file, you have to make more space in the layout for the additional rows and columns. The layout gives four character positions for each column in the weight matrix, so if you add three new hidden units, you have to add spaces to move everything that is to the right of the weight matrix further to the right by 12 columns. Also, for each added hidden unit, you have to add an additional line in the layout. These extra lines in the template file go just above the last line that has $'s on it. Both these things are done by adding blank lines and spaces to the layout. You must also alter the template specifications themselves, specifying the correct starting element and number of elements for the vector displays and specifying the correct starting row, number of rows, starting column, and number of columns for the weights.

Ex. 5.3. Other Problems for Back Propagation

Construct a different problem to study, either choosing from those discussed in *PDP:8* or choosing a problem of your own. Set up the appropriate network, template, pattern, and start-up files, and experiment with using back propagation to learn how to solve your problem.

Q.5.3.1. Describe the problem you have chosen, and why you find it interesting. Explain the network architecture that you have selected for the problem and the set of training patterns that you have used. Describe the results of your learning experiments. Evaluate the back propagation method for learning and explain your feelings as to its adequacy, drawing on the results you have obtained in this experiment and any other observations you have made from the readings or from this exercise.

Hints. Try not to be too ambitious. You will learn a lot even if you limit yourself to a network with 10 units. For an easy way out, you could choose to study the 4-2-4 encoder problem described in *PDP:8*. The files *424.tem, 424.str, 424.net,* and *424.pat* are already set up for this problem.

EXTENSIONS OF BACK PROPAGATION

Up to this point we have discussed the use of back propagation in one-pass, feedforward networks only. Here we discuss three extensions. The first maintains the strictly feedforward character of the networks we have been considering but provides for gradual build-up of activation. The second simulates recurrent activation in a limited way by using linked connections. The third introduces real recurrence of activation by allowing the units in the network to send their outputs back to specified input units. We will briefly consider each of these extensions and offer an exercise to go with each.

CASCADED FEEDFORWARD NETWORKS

One of the precursors of our work on PDP models was a model called the *cascade* model (McClelland, 1979). This model was a purely linear, feedforward, multilayer network. Units at each level took on activations based on inputs from the preceding level according to the following equation:

$$a_{ir}(t) = k_r \sum_j w_{ij} a_{js}(t) + (1 - k_r) a_{ir}(t-1) \tag{2}$$

Here r and s index some receiving level and the level sending to it, i and j index units within levels, and k_r is a rate constant governing the rate at which the activations of units at level r reach the value that the summed input is driving them toward.

Such a system has interesting temporal properties and provides a useful framework for accounting for many aspects of reaction time data (see McClelland, 1979, for details), but its computational capabilities are seriously limited. In particular, with linear units, a multilayer system can always be replaced by a single layer, and as we have noted repeatedly, there are limits on the computations that can be performed in a single layer. To overcome these limitations, multiple layers of units, with some form of nonlinear activation function, are required.

The idea we describe in this section preserves the desirable computational characteristics of the nonlinear networks we have been considering in this chapter but combines with these the gradual build-up of activation characteristic of the cascade model. To achieve this, we introduce one change into the cascade equation: Instead of directly setting the activation of units on the basis of this equation, we use it to determine the net input; the activation is then calculated from the net input using the logistic function. Thus the equation for the net input to a unit becomes

$$net_{ir}(t) = k_r \sum_j w_{ij} a_{js}(t) + (1 - k_r) net_{ir}(t-1), \tag{3}$$

and the equation for its activation is

$$a_{ir}(t) = \frac{1}{1 + e^{-net_{ir}(t)}}. \tag{4}$$

In our implementation of this scheme, there is a single rate parameter called *crate* (for cascade rate) that is used for all units rather than a separate rate parameter for each level.

One very nice feature of this scheme is that if an input pattern comes on at time $t = 0$ and stays on, the *asymptotic* activation each unit reaches is identical to the activation that it reaches in a single step in the standard, one-pass feedforward computation used up to now in back propagation.

Thus, we can view the one-pass feedforward computation as one that computes the asymptotic activations that would result from a process that may in real systems be gradual and continuous. We can then train the network to achieve a particular asymptotic activation value for each of several patterns and then observe its dynamical properties.

Before we actually turn to an exercise in which we do just what we have described, we mention one more characteristic of these cascaded feedforward networks: Their dynamical properties depend on the initial state. Here we assume that the initial state is the pattern of activation that results from processing an input pattern consisting of all-zero activations for all of the input units. For this to work, the network must in general be trained to produce some appropriate output state for this initial input state. Here we simply assume that the desired initial output state is also all-zero activations on all of the output units.

Ex. 5.4. A Cascaded XOR Network

After training the XOR network of Ex. 5.1 in the standard way, turn on *cascade* mode and, using the *test* command, examine the time course of activation of the hidden and output units for the patterns in the XOR pattern set.

To run this exercise, run the program with the *cas.tem* and *cas.str* files. These differ only slightly from the standard XOR files. The *.tem* file displays the cycle number (which is 0 until *cascade* mode is turned on). The *.str* file sets *nepochs* to 300 and sets *ncycles* to 100. After start-up, issue the *strain* command, which will cause the network to learn the solution to the XOR problem from the first exercise, finishing at epoch 289, where the *tss* falls below 0.04. Then enter *set/ mode/ cascade 1* to turn on *cascade* mode. In this mode, the activations of the units are initialized to the asymptotic values they would have for the input (0 0), which, conveniently, was in the training set with target output (0). Now use the *test* command to test each of the three nonzero patterns (patterns 1, 2, and 3). The *test* command will automatically set *stepsize* to *cycle* and turn on *single* mode, so you will be able to study the time course of activation step by step. To obtain a simple graph of the results after the run is over, you can use the *log* command to open a log file before you begin testing. Then you can use the *plot* program described in Appendix D. Once a log file has been opened with the *log* command, the *.str* and *.tem* files set up the *dlevels* of the templates and the global *slevel* of the program so that the pattern name, the cycle number, and the activations of the hidden and output units are logged on each cycle.

Q.5.4.1. Explain why the output unit is initially more strongly activated by the (1 1) input pattern than by either the (1 0) or the (0 1) patterns, and explain why the activation eventually "turns around" for the (1 1) pattern, following an overall U-shaped activation curve.

RECURRENT NETWORKS

In *PDP:8*, it is noted that it is possible to emulate a recurrent network in which units feed activation to themselves and to units that project to them by essentially duplicating each unit. Here we consider a simple three-unit example of such a network.

Conceptually, the network we will consider consists of three units, with each unit having a modifiable bias term and a modifiable connection to itself and to each other unit. Actually, though, the network consists of three layers of three units each. Each unit in the first layer projects to each unit in the second layer, and each unit in the second layer projects to each unit in the third layer. Units in the second and third layer also have modifiable bias terms. Connections from the first layer to the second layer are linked to the corresponding connections from the second to the third layer. Similarly, the bias terms of the units in the second layer are linked to the corresponding bias terms in the third layer. These links force corresponding weights and biases to be equal, so there is really only one set of weights and one set of biases, as there would be in our conceptual network. This three-layer network allows us to supply an initial input pattern to the input units and to simulate two time steps of the activation process that would occur in our conceptual three-unit network.

Ex. 5.5. The Shift Register

In this exercise we will train the three-unit recurrent network to be a shift register. The task is to learn to take an arbitrary pattern of 1s and 0s on the three input units and shift that pattern two bits to the right. This is done by training the network with input-output pairs, each consisting of one of the eight 3-bit binary input vectors paired with the same vector shifted two bits right. Actually, it is appropriate to think of the units as forming a ring so that when a bit is shifted from the right-most position it appears in the left-most position.

The exercise is carried out by calling the **bp** program with the files *rec.tem* and *rec.str*. The display shows the activation of each of the three copies of the three units, along with standard information such as the *tss*, and so on. The eight different binary patterns of three bits have been read in as input patterns, along with the same patterns shifted two bits to the right as the corresponding targets. If you run a *tall* at this point, you will see that the output of the third stage of processing is close to 0.5 for all eight of the different input patterns.

The weights and bias terms have been set up as already described, and the bias terms are additionally constrained to be negative. The other parameters were chosen arbitrarily. The problem is an easy one to learn, as you can prove to yourself by running the *strain* command. The network will generally find a solution in about 40 to 200 epochs. Once the solution has been found, you can run a *tall* and see how the network has solved the problem. If you wish to see the weights, biases, and target patterns in addition to the activation vectors, you can set *dlevel* to 3. In this state, the display will show the weight matrix twice, once with the initial set of activations feeding into it from above with the intermediate set to the left and once with the second set feeding into it with the third set to its left.

Q.5.5.1. Does the network always find the same solution to the problem? Describe all the different solutions you get after repeated runs, using *newstart* to reinitialize the network between runs. What happens if you try removing the constraints that force the biases to be negative and link the weights together? Do these changes increase the number of different solutions?

Hints. To remove the negative constraints, you simply have to set up a new *.net* file, deleting the word *negative* wherever it appears in the *rec.net* file. To remove the links, you can replace the letters *a* to *i* in the *network* specification and *j* to *l* in the *biases* specification with *r*.

SEQUENTIAL NETWORKS

A somewhat different type of recurrent network—here called a sequential network—has been proposed by Jordan (1986). He devised this type of network originally for generating a sequence of outputs, such as phonemes, as would occur, for example, in saying a word.

In Jordan's formulation, the network consists of a set of input units called *plan* units and another set of input units called *current-state* units. Both sets feed into a set of hidden units, which in turn feed into a set of output, or *next-state*, units. This type of network is illustrated in Figure 11. At the beginning of a sequence, a plan pattern is input to the plan units, and the current-state units are set to 0. Feedforward processing occurs, as in standard back propagation, producing the first output pattern (e.g., the first phoneme of the word specified in the plan). This output pattern is then copied back to the current-state units for the next feedforward sweep. Actually, the activation of each current-state unit at the beginning of the next processing sweep is set equal to the activation of the corresponding

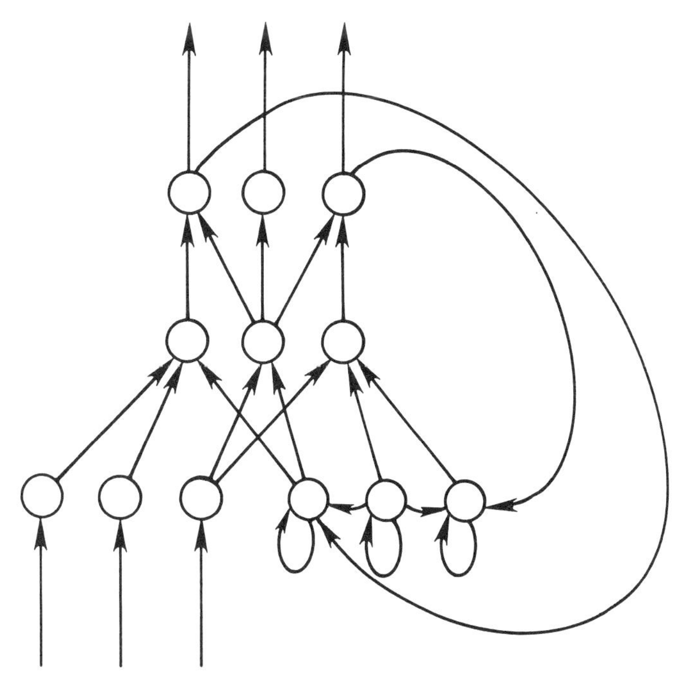

FIGURE 11. Basic structure of Jordan's sequential networks. (From "Attractor Dynamics and Parallelism in a Connectionist Sequential Machine" by M. I. Jordan, 1986, *Proceedings of the Eighth Annual Conference of the Cognitive Science Society*. Hillsdale, NJ: Lawrence Erlbaum Associates. Copyright 1986 by M. I. Jordan. Reprinted by permission.)

output unit plus μ (*mu*) times the unit's previous activation. This gives the current-state units a way of capturing the prior history of activation in the network.

We have provided a simple facility for implementing Jordan's sequential nets as well as many more complex variants. This facility employs negative integers in input patterns to indicate that the unit's activation should be derived from activations produced on the previous cycle. The particular negative integer indicates which unit's activation should be used in setting the activation of the input unit. Thus, if element i of a particular input pattern is equal to $-j$, the activation of unit i will be set to $a_j + \mu a_i$, where a_i and a_j are taken to be the activations of units i and j from the previous cycle. The parameter *mu* can be set using the *set/ param/ mu* command (its value should be between 0 and 1).

To implement Jordan's sequential networks using this scheme, we first set up a network, say, with four plan units, four current-state units, four hidden units, and four next-state units. Then we set up each sequence we want to train the network to learn as a set of pattern pairs. In each pattern pair belonging to the same sequence, the plan field stays the same. The target patterns are the sequence of desired outputs. For the first pattern in the sequence, the input values for the current state units are 0. For subsequent patterns in the sequence, the input values for the current state units are set to $-n$, where n is the index of the corresponding output unit. Thus, the patterns to specify that the network should interpret the plan pattern (1 0 1 0) as an instruction to turn on first the first output unit, then the second, then the third, and then the fourth would be as follows:

```
Plan          Current State        Target
1 0 1 0        0   0   0   0       1 0 0 0
1 0 1 0      -12 -13 -14 -15       0 1 0 0
1 0 1 0      -12 -13 -14 -15       0 0 1 0
1 0 1 0      -12 -13 -14 -15       0 0 0 1
```

(In the pattern file each pattern would also have a name at the beginning of the line.) Note that the assumptions made by Jordan can be seen as a special case of a very general class of sequential models made possible by the facility to set the activation of input units based on prior activations. Inputs can be set to have activations based on the activations of hidden units as well as output units, and external input can be intermixed at will with feedback. It is even possible to set up any desired combination of previous inputs by hard-wiring hidden units to receive inputs from particular sets of input units.[6]

Ex. 5.6. Plan-Dependent Sequence Generation

To illustrate the use of sequential networks, we provide a simple exercise in which a back propagation network is trained to turn on the output units in sequence, either left to right if the pattern over the plan units is (1 0 1 0) or right to left if the pattern is (0 1 0 1). As in the example just described, the network has four plan units, four current-state units, four hidden units, and four output, or next-state, units. The network is trained with the four patterns from each of the two desired sequences, for a total of eight training patterns.

To run this exercise, start up the **bp** program with the files *seq.tem* and *seq.str*. The activations of the plan units and current-state units will be shown in a vertical column on the left of the screen, the activations of the hidden units to the right of these, and the activations of the output units to the right of these. Before training, run *tall* to see that the output is weak and random, then run *strain* until the network solves the problem (it

[6] The only restriction on the use of the previous-state facility is that an input unit cannot receive its input from the previous activation of a lower-numbered unit. In practice this should not be much of a restriction since lower-numbered units would always be other input units.

generally takes several hundred epochs). Once the problem is solved (*tss* less that 0.1), do a *tall* again to verify that the network does as instructed.

Q.5.6.1. See if your can figure out some of the reasons why this task is harder than the XOR task. If you want to explore this kind of network further, you might want to study the effects of various parameters on learning time. For example, you might want to experiment with different values of *mu*, *lrate*, and *momentum*.

Hints. In thinking about the first part of the question, you might note what happens to the patterns of activation on the output units, and therefore the current-state units, during the early parts of training. In your explorations for the second part of the question, the most interesting effects occur with variations in *mu*. See if you can discover and understand these effects.

CHAPTER 6

Other Learning Models: Auto-Associators and Competitive Learning

There are a number of learning paradigms in PDP systems—each with a characteristic goal or task. These paradigms include the pattern association paradigm, in which the goal is to learn mappings between specific input and output pairs; the auto-associator paradigm, in which the goal is to store specific patterns for future retrieval; and the regularity detection paradigm, in which the goal is to discover salient features of the ensemble of patterns. Thus far in this book we have focused almost entirely on the pattern association paradigm for learning. Clearly the pattern associator of Chapter 4 and the back propagation model of Chapter 5 are both examples of systems learning input-output mappings. The current chapter focuses on the other two paradigms. We begin with a discussion of several simple auto-associators and then move to a discussion of one of the most studied regularity detection models, *competitive learning.*

THE AUTO-ASSOCIATOR

BACKGROUND

The auto-associator models are a class of related models that share the auto-associative architecture. That is, they all consist of a single set of units that are completely interconnected. In some ways, this architecture is the most general architecture for a connectionist system; all other architectures are more restricted subsets of this architecture. However, given the learning rules that we will be exploring for training these networks in the present chapter, auto-associators are limited by the fact that they can only train connections between units whose target activations can be specified from outside the network.

In spite of this limitation, auto-associators have several interesting properties. They can learn to do pattern completion and to rectify or restore distorted versions of learned patterns to their original form. They can learn to extract the prototype of a set of patterns from distorted exemplars presented during training. Discussions of these and other aspects of auto-associators may be found in Anderson (1977), Anderson, Silverstein, Ritz, and Jones (1977), Kohonen (1977), and in *PDP:17* and *PDP:25*.

The auto-associator models we will consider in this section are similar to pattern associators, with one major difference: There is only a single set of units, and instead of having connections from input units to output units, each unit serves as both an input unit and an output unit, so that each unit is connected to every other unit. In some versions, it may also be connected to itself. A picture of an auto-associator is shown in Figure 1. In all the versions of the auto-associator that we will consider here, input patterns consist of vectors specifying positive and negative inputs to the units from outside the network. There are no bias terms on the units. Units take on activation values that may be positive or negative, based on these external inputs and on the connections they receive from other units inside the network.

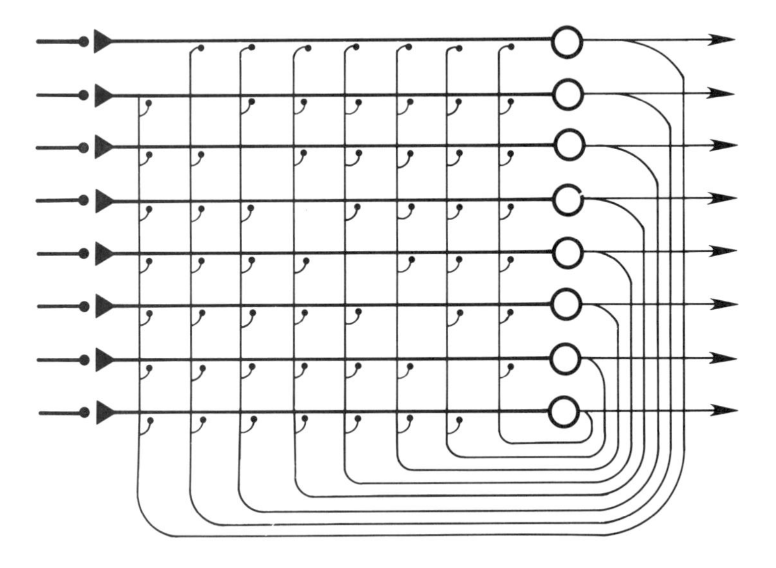

FIGURE 1. A simple eight-unit auto-associative network. (From "Distributed Memory and the Representation of General and Specific Information" by J. L. McClelland and D. E. Rumelhart, 1985, *Journal of Experimental Psychology*, *114*, 159-188. Copyright 1985 by the American Psychological Association. Reprinted by permission.)

A basic understanding of the essential properties of the auto-associator can best be achieved by considering a linear, Hebbian version of a pattern associator in which the input patterns and the output patterns happen to be the same. For this case we can use what we learned in Chapter 4 about the pattern associator, noting that the associations are now between a pattern and itself. Specifically, we recall that the output produced in response to test input pattern $\mathbf{i}_t$ is proportional to the sum of the output patterns experienced during learning, each weighted by the similarity of the corresponding input pattern to the test input pattern:

$$\mathbf{o}_t = k\sum_l \mathbf{o}_l(\mathbf{i}_l \cdot \mathbf{i}_t)_n. \quad (1)$$

The constant of proportionality, k, is equal to the learning rate parameter, ϵ, times the number of units in the network. Since we are considering the case in which the training consists of associating each input vector with itself, the training output vectors $\mathbf{o}_l$ can be replaced with the training input vectors $\mathbf{i}_l$. In this case, the output at test is equal to the sum of the input patterns used during training, each weighted by its similarity to the input pattern used at test. Given this equation, we can immediately observe the following points:

- If a test input pattern is orthogonal to all of the input patterns used during training, then the network will produce a null output.

- If a test pattern is orthogonal to all but one of the training patterns and is identical to this other training pattern, then the output will be equal to the test pattern scaled by the value of k.

- If the same pattern is presented m times during learning, then it will be as though there are m patterns "stored" in the network that are identical to it. Therefore if this same pattern is presented as a test input, the output will be equal to m times k times the test pattern.

More succinctly, we can say that if we associate a set of orthogonal patterns, each with itself, in a linear Hebbian associator, and if we test with one of these stored patterns, then the output will be equal to a scaled version of the input, and the scale factor will be proportional to the number of times we have experienced the pattern during learning.

Patterns that are scaled by a network are called *eigenvectors*; eigenvector simply means "same vector." The magnitude of the eigenvector, as it is processed by the network, is called its *eigenvalue*. For our linear Hebbian auto-associator trained with an orthogonal set of learning patterns, the learned patterns form a set of eigenvectors. Their eigenvalues are km_l, where m_l is the number of presentations of learning pattern l.

Now, however, suppose that we present a pattern that has some similarity to each of several different stored patterns. Then we find that the output produced is a blend of these stored patterns, with the contribution of each weighted by its similarity to the test pattern times its eigenvalue. As an example, suppose we have stored these two patterns:

$\mathbf{i}_0$: +1.00 +1.00 +1.00 +1.00 −1.00 −1.00 −1.00 −1.00
$\mathbf{i}_1$: +1.00 −1.00 +1.00 −1.00 +1.00 −1.00 +1.00 −1.00

and we test with the following pattern:

$\mathbf{i}_t$: +1.00 +1.00 +1.00 +1.00 −1.00 −1.00 +1.00 −1.00

We find that the normalized dot product of pattern $\mathbf{i}_0$ with pattern $\mathbf{i}_t$ $(\mathbf{i}_0 \cdot \mathbf{i}_t)_n$ is 0.75, and the normalized dot product of pattern $\mathbf{i}_1$ with pattern $\mathbf{i}_t$ $(\mathbf{i}_1 \cdot \mathbf{i}_t)_n$ is 0.25. If each has been stored exactly once and k is equal to 1.0, then we will get as our output 0.75 times $\mathbf{i}_1$ plus 0.25 times $\mathbf{i}_2$, so the resulting output pattern is

$\mathbf{o}_t$: +1.00 +0.50 +1.00 +0.50 −0.50 +1.00 −0.50 −1.00

This vector is not the same as the input vector $\mathbf{i}_t$, so $\mathbf{i}_t$ is not an eigenvector of this network. The response is a weighted sum of the stored vectors, with the weights depending both on the similarity of the input to each stored vector and on the eigenvalues of these vectors. We will see in the exercises that when the output of the auto-associator is fed back into itself and nonlinearities are introduced, the output can often end up exactly matching the most similar pattern used during learning. We call this process the *pattern rectification process*.

A special case of pattern rectification is what is called the *pattern completion process*. This is what happens when we present an incomplete vector in which some of the +1s and −1s have been replaced by 0s. Thus, if we have previously stored patterns $\mathbf{i}_0$ and $\mathbf{i}_1$ as above, we can present an incomplete version of one of these patterns as a test input pattern and the network will fill in or complete the remainder. Thus suppose we present the following test pattern:

$\mathbf{i}_t$: +1.00 −1.00 +1.00 −1.00 0.00 0.00 0.00 0.00

In this case, $\mathbf{i}_0 \cdot \mathbf{i}_t$ is 0.0 and $\mathbf{i}_1 \cdot \mathbf{i}_t$ is 0.5. The network will produce the output vector $\mathbf{o}_t$,

$\mathbf{o}_t$: +0.50 −0.50 +0.50 −0.50 +0.50 −0.50 +0.50 −0.50

in response to this input. Note that this vector points in the same direction as the stored vector $\mathbf{i}_1$, but it is of lesser magnitude.

In general, in completion with orthogonal input patterns and linear units we obtain a scaled version of the incomplete stored vector that is probed, where the scale factor is equal to the normalized dot product of the stored vector and the incomplete version of it that is used as the probe.

The pattern completion and rectification processes we have been describing are general characteristics of auto-associator models. Another general characteristic is their tendency to learn to respond better to the prototype, or central tendency, of a set of distorted exemplars of a category than to any of the individual distortions themselves. This characteristic arises from the fact that each new distortion learned is superimposed in the connection strengths; the characteristics of the individual exemplars tend to average out as more and more exemplars are presented. This characteristic of auto-associators is discussed at length in Anderson et al. (1977) and in *PDP:17*, and is explored extensively in the exercises.

So far we have been treating the auto-associator as if it were a pattern associator in which the input and output patterns just happen to be the same. In fact, though, the input and output patterns happen to be the same because the input and output units are really the same units. This gives the auto-associator the capability of multiple processing cycles in which the initial pattern of activation is produced by some external input, and each successive cycle involves updating the activations of the units, based on the continuing external input, plus what we call the *internal input*—the input to each unit via the connections internal to the net. The internal input to unit i, $intinput_i$, is given by

$$intinput_i = \sum_j w_{ij} a_j .$$

This internal input is equivalent to the output that would be produced by a linear pattern associator. In the auto-associator, it is combined with the continuing external input to each unit, and is then treated in different ways in the different variants of the auto-associator model, which are described later.

Learning Regimes for Auto-Associators

Both the Hebb rule and the delta rule are available for use in auto-associator models. When the Hebb rule is used, the external input is assumed to be clamped onto the units for the purpose of training. In this case the formula for updating the weights is

$$\Delta w_{ij} = \epsilon (extinput_i)(extinput_j).$$

When the delta rule is used, the external input pattern is applied at the beginning of time cycle 1 and is left on. Processing goes on for *ncycles*. At the end of *ncycles*, a variant of the delta rule is used to adjust the strengths of the connections in the network. In this variant, the goal of learning is to have the internal input to each unit match the external input. In this case, the error measure for each unit, $error_i$, is defined to be

$$error_i = extinput_i - intinput_i$$

where the $intinput_i$ is the value at the end of *ncycles* of processing, based on the activations at the end of the preceding cycle.

In the general formulation of the auto-associator, each unit is assumed to be connected to every other unit, including itself. In networks with large numbers of units, these self-connections are unimportant, but in smaller networks trained with the delta rule, where the goal is to learn connections that foster pattern completion and rectification, strong self-connections can tend to defeat learning. This is because self-connections allow units to predict their own activation, thus reducing the error and preventing the network from learning strong between-unit connections that can perform the completion and rectification processes. Thus, when the delta rule is used in an auto-associator, it is best to force the connection from each unit to itself to remain fixed at 0.

Limitations of the Auto-Associator

The limitations of the auto-associator are similar to the limitations of the pattern associator. When trained using the Hebb rule, perfect reproduction of learned patterns can only be obtained if orthogonal patterns are used; with nonorthogonal patterns there is always some cross-talk between the patterns. When trained using the delta rule, the learning process converges only if the following linear predictability constraint can be met:

> Over the entire set of patterns, the external input to each unit must be predictable from a linear combination of the activations of each unit that projects to it.

This constraint, for example, prevents the auto-associator without hidden units from learning to turn on a unit when two other units are both on or both off, while at the same time turning the unit off when one of the two other units is on and one is off.

Auto-associators can be constructed in which there are hidden units, of course; a simple example is described at the end of *PDP:17*. More generally, encoder networks as described in *PDP:5* are examples of auto-associators with hidden units. The auto-associator models used in the present chapter, however, do not contain hidden units.

In the sections that follow, we describe three main variants of the auto-associator. All of these will be considered in the exercises.

The Linear Auto-Associator

Perhaps the simplest variant of the auto-associator is what we will call the linear auto-associator. In this model, the change in activation of each unit on each processing cycle is a weighted sum of the external and internal inputs to the unit, less a decay term that tends to restore activation to a resting level of 0:

$$\Delta a_i = (estr)extinput_i + (istr)intinput_i - (decay)a_i . \qquad (2)$$

Note that the $extinput_i$ and $intinput_i$ together make up the net input to unit i (there is no bias term). The parameters *estr* and *istr* scale the contributions of the external and internal input to each unit, as in the constraint satisfaction models considered in Chapter 3.

This model is mathematically very simple, and it is typically used in the following way. At some time $t=0$, activations of all units are set to 0. At the beginning of cycle 1, a pattern of +1s and −1s is supplied as the external input and is left on until the end of *ncycles* of processing. On the first cycle of processing, since the prior activations of all the units are all 0, each unit takes on an activation equal to *estr* times the external input pattern. After that, processing proceeds in accordance with Equation 2. For simplicity, we will study the case in which the decay parameter is set to 1.0. In this case, the activation of each unit at time t ($a_i(t)$) is given by

$$a_i(t) = (estr)extinput_i(t) + (istr)intinput_i(t).$$

Here $intinput_i(t)$ is based on the activations of the units at time $t-1$.

A Difficulty with the Linear Auto-Associator

The linear auto-associator model is very useful for illustrating the basic pattern completion and regularization processes described above. A difficulty, however, is that the network can "blow up"; that is, activations can become very large, very quickly as a result of the self-reinforcing feedback characteristic of the network. When the model is run with *ncycles* equal to 2, this is not a problem. With larger values of *ncycles*, some form of nonlinearity must be introduced. The next two variants of the auto-associator involve introducing different types of nonlinearity into the basic model.

The Brain-State-in-the-Box Model

One form of nonlinearity that keeps activations from growing without bound is introduced in the "brain state in the box" or BSB model proposed by Anderson et al. (1977). In this model, activations are prevented from growing larger than $+C$ or smaller than $-C$. In our version of this model, we will use $C = 1.0$.

The effect of this "clipping" operation, of course, is to prevent activations of units from growing without bound; instead it keeps them in a hypercube, or box, bounded by +1.0 and −1.0 on each dimension. Small inputs may still be amplified by the network, but when the activations of the units reach +1.0 or −1.0, they are simply cut off. This has an interesting side effect: It means that processing tends to result in patterns of activation that correspond to corners of the hypercube, that is, states that consist of all +1s and −1s. The corners tend to correspond to the patterns that had previously been learned. In this case, as we shall see in the exercises, the auto-associative process tends to drive incomplete or distorted versions of stored patterns toward the stored patterns, producing perfect rectification and completion.

In the version of the BSB model that we shall consider in the exercises, the learning rule is the same as in the linear model already described. It is also possible to use the variant of the delta rule described earlier with the BSB model.

The DMA Model

The final auto-associator model we will consider is the model of distributed memory and amnesia described in *PDP:17* and *PDP:25*. Here we call this model the *DMA model.* This model grew out of our work with the interactive activation and competition scheme described in Chapter 2. In this model, we think of the combined external and internal input to each unit as driving the activation of the unit upward or downward, depending on whether it is excitatory or inhibitory. The magnitude of the effect of the input is dependent on the distance to the maximum or minimum activation value. First we define the net input to unit i:

$$netinput_i = (estr)extinput_i + (istr)intinput_i.$$

If the net input is positive,

$$\Delta a_i = netinput(max - a_i) - (decay)a_i,$$

and, if it is negative,

$$\Delta a_i = netinput(a_i - min) - (decay)a_i.$$

This model is similar to the BSB model in that activations are kept between the values of *max* and *min*, which are set to +1.0 and −1.0. The main difference is that activations always level off at less extreme values, since at some point the "restoring" force of the decay term will match the "perturbing" force of the net input term.

Learning in the DMA model takes place using the variant of the delta rule we described earlier. In this rule, when the error is 0, the internal input to a unit matches the external input, and the total net input to a unit is the sum of the *istr* and *estr* parameters times the external input:

$$netinput_i = (estr + istr)extinput_i .$$

In contrast, before learning, when the internal input is 0, we find that the net input is simply

$$netinput_i = (estr)extinput_i .$$

The effect of this difference is to change the asymptotic activation values of the units. From our consideration of the interactive activation and competition model of Chapter 2, we recall that at asymptote, the activation of a unit is given by

$$a_i = \frac{netinput_i}{netinput_i + decay}.$$

Before learning, then,

$$a_i = \frac{(estr)extinput_i}{(estr)extinput_i + decay},$$

while after learning,

$$a_i = \frac{(estr + istr)extinput_i}{(estr + istr)extinput_i + decay}.$$

In most of the simulations reported in *PDP:17* and *PDP:25*, *estr*, *istr*, and *decay* were all set to 0.15, and external inputs used in training patterns are always patterns of +1s and −1s. This means that before learning, units take on activations of 0.50 times the sign of the external input; after learning, this value grows to 0.67. These values, of course, can be moved around at will by changing the values of *estr*, *istr*, and *decay*. The basic point is that the network is more strongly activated by familiar patterns than by unfamiliar ones. It also exhibits pattern completion and rectification, as in the other variants of the auto-associator.

IMPLEMENTATION

The auto-associative models are implemented in the **aa** program. In this program, processing is implemented much as it is in the **iac** program described in Chapter 2. The main difference is that in **aa** the output of a unit is identical with its activation; there is no check to see that the activation exceeds threshold. Both positive and negative activation values exert influences on other units.

What the **aa** program adds to **iac** is an outer loop that runs epochs of training trials. In each trial, after *ncycles* of processing, the error measure is computed and the connection strengths are modified. The routines for doing this are analogous to those used in the pattern associator.

RUNNING THE PROGRAM

The **aa** program is run in much the same way as the programs already described. The program is called with a *.tem* file and a *.str* file. Because of the simplicity of the **aa** architecture (each unit connected to every other unit), a *.net* file is not needed; instead, *nunits* is defined near the top of the *.str* file. This leads the program to create a network of *nunits* units, with a connection from each unit to itself and every other unit. Generally, a *.pat* file is used to specify a list of patterns for use in training and testing.

The *.str* file generally specifies the size of the net *(nunits)* and specifies which of several possible modes should be on or off. The DMA model is the default. The linear Hebbian model can be studied by setting *linear* mode to 1 and by setting the *hebb* mode to 1. You can study the BSB model by setting the *bsb* mode to 1. There is also a *selfconnect* mode, which is set to 0 by default; in this mode the weight from a unit to itself is forced to remain at 0.0. To study the effects of allowing nonzero self-connections this mode can be set to 1.

The facilities for training and testing are the same as those used in the **pa** and **bp** programs. The *strain* command is used to train the network using a fixed sequential order of training in each epoch. The *ptrain* command is used to train the network using a permuted order of presentations in each epoch. Both commands run *nepochs* of training, ending when interrupted or when the total sum of squares *tss* becomes smaller than the criterion *ecrit*.

During training, it is possible to specify that the training patterns should be randomly distorted. In **aa**, distortion is done by independently changing the sign of each bit (from + to − or from − to +) in each training pattern with probability *pflip* before it is presented to the network for training. A *pflip* of 0 produces no changing; a *pflip* of .5 produces totally random patterns. Note that this method of distortion is different from the one provided in the **pa** program.

To test the network, the *test* command allows testing using either one of the stored patterns, a distortion of one of these patterns, or any pattern entered directly as a sequence of +'s and −'s. At the end of *ncycles* of processing, the normalized dot product of the input with the output, the normalized length of the activation vector produced, and the correlation of the output with the external input are displayed. The *ctest* command is used for testing the pattern completion capability of the model. It allows the user to specify a part of a pattern to clear to see how well the model can do in filling it back in again. In this case the *ndp*, *nvl*, and *vcor* measures apply to the subpattern of activation filled in by the network on the cleared units rather than to the overall pattern of activation.

New or Altered Commands

The following list mentions only those commands in the **aa** program that are not the same as commands in the **pa** program.

ctest
: Allows the user to perform a completion test on an individual pattern. The user specifies which input units to clear, (that is, to set to 0) for completion testing. The *ctest* command prompts for a pattern name or number to test, then asks for a first element to clear (a number from 0 to *nunits* – 1), and then asks for a last element to clear (the last element must be greater than the first and less than *nunits*). Both the beginning and the end elements given are cleared, as well as all the units in between. The statistics computed (*ndp*, *nvl*, and *vcor*) will apply to the cleared portion of the pattern, assessed against the pattern that would have been present had these bits not been cleared.

test
: Allows testing of an individual pattern. The following arguments can be given:

  - *#N* Instructs *test* to use the corresponding pattern from the pattern list (*N* is a pattern name or number).
  - *?N* Instructs *test* to use a distorted version of the corresponding pattern (*N* is as above). Each element has its sign flipped with probability equal to the value of the *pflip* parameter.
  - *L* Instructs *test* to use the last pattern tested; this pattern is left in place.
  - *E* Instructs *test* to accept a pattern entered by the user. Pattern elements are floating-point numbers or ".", "+", or "–", corresponding to 0.0, +1.0, and –1.0. Elements must be separated by spaces and the list of elements must be terminated by *end* or an extra *return*.

get/ patterns
: Reads in a pattern file containing a list of pattern specifications. Each pattern specification consists of a pattern name followed by *nunits* entries indicating the values of each element of the pattern.

Entries can be floating-point numbers or "+" (for 1.0), "–" (for –1.0), or "." (for 0.0).

get/ rpatterns

Causes the program to construct a set of random patterns (vectors of +1s and –1s) with a specified probability that each unit will be +1. Prompts for two arguments as follows:

How many patterns?
(give desired number of patterns to construct)
make input + with probability:
(give desired probability for elements to be positive)

This list is stored in the program's internal *ipattern* list and can be saved using the *save/ patterns* command. Patterns are assigned names of the form *rN* where *N* is the pattern number.

save/ patterns

Allows the user to save the patterns in the program's pattern list in a file.

The **aa** program does not provide a *cycle* command to continue cycling if you wish to run more cycles with the *test* or *ctest* commands. Instead you must set *ncycles* to a larger number and run the *test* or *ctest* command again. With *test* you can enter *L* as the argument to exactly repeat the previous test.

Variables

The following list mentions only those variables that are new or different in the **aa** program. As usual, all of the variables are accessed via the *set/* and *exam/* commands.

stepsize

The default *stepsize* in **aa** is *pattern*. This means that a step consists of presenting an input pattern as the external input, resetting all the activations in the network, running *ncycles* of processing, computing error information and summary statistics, and changing weights if *lflag* is set. Other possible values of *stepsize* are *cycle*, which causes updating/pausing to occur after each cycle; *epoch*, which causes updating/pausing to occur only at the end of an entire processing epoch; and *nepochs*, which causes updating/pausing to occur only at the end of *nepochs*.

mode/ bsb

When *bsb* is set to 1, activations are clipped at +1 and –1. This mode has no effect unless the *linear* mode is also in force since activations are otherwise restricted to the [1, −1] interval by the DMA activation equations.

mode/ hebb
When *hebb* is set to 1, the program uses the Hebbian learning rule. When *hebb* is 0 (the default), the delta rule is used.

mode/ linear
By default, the activations are updated according to the DMA activation equations. When *linear* is set to 1, the activation process is linear, subject to clipping at +1 and −1 if *bsb* mode is also set.

mode/ selfconnect
By default, when *selfconnect* is 0, the weight from each unit to itself is fixed at 0.0. When *selfconnect* is set to 1, self-connections are trained just like all other connections in the network.

param/ estr
Scales the magnitude of the external input to each unit. The scaling is applied in determining the net inputs to the units but is not applied in computing errors.

param/ istr
Scales the magnitude of the internal input to each unit. Scaling is applied as with *estr*.

param/ lrate
The learning rate parameter. Generally, its value should be less than 1/*nunits*.

param/ pflip
The probability that pattern elements have their signs flipped during training and when flipping is requested in using the *test* command.

state/ error
Vector of errors for each unit. Each element is the difference between the unit's external input and its internal input.

state/ extinput
Vector of external inputs to units. Note that this is displayed before the effects of scaling the external input by the *estr* parameter are applied.

state/ intinput
Vector of internal inputs to units from other units. Note that this vector is displayed before the effects of scaling the external inputs by the *istr* parameter are applied.

state/ ndp
Normalized dot product of the current external input pattern with the current activation pattern. Updated at the end of every cycle when *stepsize* = *cycle* or at the end of every epoch otherwise.

state/ nvl
The normalized length or strength of the activation vector. Updated like *ndp*.

state/ prioract
Vector of activations from the preceding processing cycle.

state/ vcor
The vector correlation of the present pattern of activation with the external input. Updated like *ndp*.

OVERVIEW OF EXERCISES

We provide four exercises for use with the different auto-associator models. Ex. 6.1 explores the linear Hebbian associator and examines its handling of sets of orthogonal patterns. Ex. 6.2 explores the BSB model and its pattern completion and reactivation capabilities. Ex. 6.3 examines the linear auto-associator with delta rule learning, focusing on exploring the characteristics of ensembles of patterns that influence whether they can be learned in a one-layer auto-associative network. Finally, Ex. 6.4 examines some of the psychological characteristics of auto-associator models, and allows the user to run variants of several of the examples discussed in *PDP:17* (pp. 182-192).

Ex. 6.1. The Linear Hebbian Associator

In this first exercise, you can familiarize yourself with the use of the **aa** program and study the effects of learning sets of patterns in the linear Hebbian auto-associator.

To start you off, we have provided the following relevant files: *lh8.tem, lh8.str*, and *two.pat*. The *lh8.str* file sets up a linear Hebbian auto-associator with eight units. It sets *hebb* mode to 1, sets *linear* mode to 1, and sets *selfconnect* mode to 1. It sets several parameters to values that make the behavior of the auto-associator particularly transparent. The *decay* variable is set to 1.0. This means that the activation on each trial is simply the sum of the external input (which is turned on and left on throughout processing) and the internal input from the units in the network, based on the pattern of activation achieved at the end of the previous cycle of processing. The *istr* and *estr* parameters are set to 1.0, so the net input is equal to the sum of the external input plus the internal input. The *lrate* parameter is set to 1/*nunits*, or 0.125. This means that in one learning trial, weights that give each of several orthogonal patterns an eigenvalue of 1.0 will be stored in the network. For initial testing, the file *two.pat* is supplied with two orthogonal patterns named *a* and *b*.

To run the program, you type:

aa lh8.tem lh8.str

The resulting screen display is similar to the display for the **pa** program. The left column displays some of the prominent variables relevant to the Hebbian auto-associator. The first three are the pattern number, the cycle number, and the epoch number. Below these are the normalized dot product of the activation vector with the external input, the normalized length of the pattern of activation, and the correlation of the pattern of activation with the external input vector. To the right of these variables is the weight matrix. This matrix shows the value of the weight from the unit in each column to the unit in each row. These values are multiplied by 100, so 10 means 0.10 and 100 means 1.0. Of course, reverse video indicates negative numbers as in other programs. To the right of the weights are the external input pattern, the internal input pattern, and the activation pattern that results from these inputs. All these are scaled by 100 as well, so that 100 stands for an actual value of 1.0. Below the weight matrix, the *prioract* vector is displayed. This vector represents the pattern of activation that was present at the end of the previous processing cycle. Like the *activation* vector, this vector is initialized to 0.0 at the beginning of processing each input pattern.

The display shown in Figure 2 shows the results of the first cycle of processing pattern *a* from the file *two.pat*. The file *two.pat* was read in by entering *get/ pat/ two.pat*, and then *single* was set to 1. Following this, the command *strain* was entered. This command runs *nepochs* of learning, but for the example *nepochs* is set to 1, so each pattern will be presented for learning only once as a result of entering this command at this point. After the *strain* command was entered, the program set the external input to

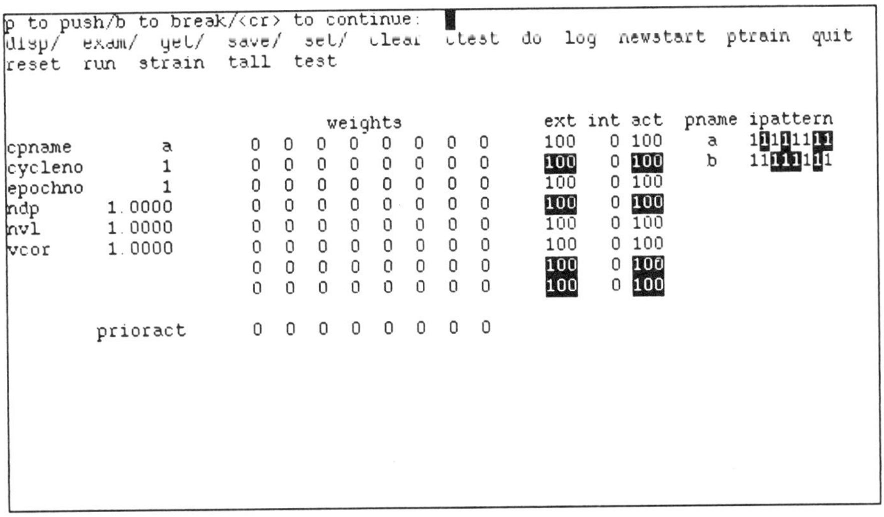

```
p to push/b to break/<cr> to continue:
disp/  exam/  get/  save/  set/  clear  ctest  do  log  newstart  ptrain  quit
reset  run  strain  tall  test

                              weights                     ext int act  pname ipattern
cpname        a      0  0  0  0  0  0  0  0       100   0 100    a    11111111
cycleno       1      0  0  0  0  0  0  0  0       100   0 100    b    11111111
epochno       1      0  0  0  0  0  0  0  0       100   0 100
ndp      1.0000      0  0  0  0  0  0  0  0       100   0 100
nvl      1.0000      0  0  0  0  0  0  0  0       100   0 100
vcor     1.0000      0  0  0  0  0  0  0  0       100   0 100
                     0  0  0  0  0  0  0  0       100   0 100
                     0  0  0  0  0  0  0  0       100   0 100

       prioract      0  0  0  0  0  0  0  0
```

FIGURE 2. The display produced by **aa** with an eight-unit network while processing an input pattern before any learning has taken place.

equal pattern *a* (the vector +−+−++−−) during the first cycle of processing and set the activations of the units based on these external inputs, then paused with the display shown in Figure 2.

The display indicates that the weights are all 0s, that the external input is a pattern of +1s and −1s, matching the first input pattern, that there is no internal input, and that the activations of the units are +1 and −1, matching the external input. The activations are equal to the external inputs since the network is linear and *estr* is equal to 1. So far no internal input has been generated.

To run another cycle, simply type *return*. In this case nothing changes: the external input is still the only thing influencing the activations of the units because the weights are all 0. Since *ncycles* is set to 2, this is the end of processing the first pattern.

At this point, another *return* results in the first pattern being stored in the weights using the Hebb rule. The display that is presented at this point reflects the new values of the weights, as well as the external input that gave rise to them, and the pattern of activation at the end of the preceding processing cycle. Note that the value of the learning rate parameter at this point is 1/*nunits*, or 0.125, as specified in the *lh8.str* file.

Q.6.1.1. Make sure you understand the weight matrix. First, be sure you know which is the receiving unit and which is the sending unit for the weight shown in row 3, column 0. Look at the weights in row 3 of the weight matrix. Why do they have these values? Be sure to explain both the sign and magnitude of the weights. (Remember that the values of the weights are displayed as hundredths and are truncated to only two places, so 12 corresponds to 0.125.)

When the next *return* is entered, the activations are cleared, the second pattern is presented so that its values now appear on the external input, and the first processing cycle is run. At this point the activations reflect the external input alone because there has not yet been a chance for activation to propagate through the connections. After the next *return*, however, the pattern of activation at the end of cycle 1 has had a chance to generate excitatory and inhibitory influences on other units by way of the connections in the network. The reader will note, however, that the internal input to each unit is still 0 at this point.

Q.6.1.2. Explain why the internal input to each unit is 0, even though each unit is producing a nonzero contribution to the net input of every unit.

After another *return*, the second pattern is stored in the weights, and the resulting matrix of weights is displayed. Type one more *return* to get back to the **aa:** prompt.

Q.6.1.3. Describe and explain the weights in row 2 of the weight matrix.

You have now completed training the network once with each of the two patterns in the file *two.pat*, and you are ready to see what happens when you test these two patterns. To do this you should use the *tall* command: when this command is executed, no learning occurs. The patterns are presented in sequential order and processed just as before, but there are no changes made to the weights.

Q.6.1.4. What happens when each of the two learned patterns is processed? Explain.

Now you are ready to try a training experiment of your own. Using the file *two.pat* as your model, generate your own set of four orthogonal patterns; include in the set the two patterns in *two.pat*. Read it into the network using the *get/ patterns* command. (We supply a file called *four.pat* that can be used, but it is better to make up your own.) Test all four patterns using *tall* (see Q.6.1.5 below), based on the weights obtained by training with the patterns in *two.pat*.

You can use the *strain* command to train the network with this new set of patterns on top of the connection strengths obtained by training on the *two.pat* patterns. The new changes to the weights will be added to the changes that are already in place from the first training set. This means that two of the patterns will have been learned twice, whereas the other two will have been learned once each.

Q.6.1.5. Display your set of four orthogonal patterns. Explain what happens when you test each of these four patterns, both before and after learning. Also, describe and explain what happens when you test the network with a vector that is equal to -1.0 times one of the stored vectors. Refer to the facts about eigenvectors in your explanation.

Hints. You can use the *test* command to enter the vector for this last part of the question.

Q.6.1.6. Set the number of processing cycles (*ncycles*) to 4, and use the *test* command to test one of the new (once-learned) patterns and one of the old (twice-learned) patterns. What happens with each? Explain.

Now construct a set of eight orthogonal patterns of +1s and −1s (hint: one of the vectors must be all +1s or all −1s). Reset the weights to 0 using the *reset* command, set *ncycles* back to 2, and train the network through one training epoch on all eight patterns, using the *strain* command.

Q.6.1.7. Describe the set of weights that results from this training experience. What will happen at this point if you present an arbitrary vector of +1s and −1s? Explain both in terms of the weights in the network and in terms of the eigenvectors of the network.

Ex. 6.2. The Brain State in the Box

One of the flaws of the linear Hebbian associator is that it can "blow up," as you will have seen in Ex. 6.1. You can overcome this limitation, however, by using the brain-state-in-the-box model; that is, by simply stipulating that units have maximum and minimum activations that cannot be exceeded. Our implementation arbitrarily sets these as +1.0 and −1.0. You can implement this model by using the following command:

aa: *set mode bsb 1*

Under these circumstances, it is more interesting to start with weaker external inputs so that they have some room to grow before they are clipped off at the corners. You can make the external inputs weaker by using the *estr* parameter. For example, setting it to 0.1 will mean that an external input specified as having a magnitude of 1.0 will actually only add 0.1 to the net input of the unit receiving it. Note that the *extinput* column in the display gives the external input specification *before* multiplication by *estr*.

The file *bsb8.str* turns on *bsb* and sets *estr* to 0.1, so you can set up for this exercise by restarting the program with the command:

aa lh8.tem bsb8.str

Completion and rectification. In this part of the exercise, you will see how the BSB model's corner-seeking characteristics lead to the pattern completion and rectification capabilities of this type of auto-associative network. First, train your BSB network with the two patterns in the file *two.pat*. Run the network for two training epochs. (It's easiest to just run the *strain* command twice.) At this point the weights should have values of +0.5, −0.5, and 0.0. You may notice what looks like an anomaly; the external activation is shown as +1.0 (100) and −1.0 (reverse video 100), but the internal input is ±0.1, and the activation is only ±0.2. This is because the *estr* parameter is set to 0.1. Thus, external inputs with an absolute value of 1.0 only add 0.1 to the net input.

At this point, you should set *ncycles* to 8. Now you're ready to do the following tests.

Q.6.2.1. Use the *tall* command to test the two learned patterns. Describe and explain the time course of build-up of activation of the units.

Explain in terms of the eigenvalues of the stored patterns and in terms of the BSB activation assumptions.

Q.6.2.2. Use the *ctest* command to present the pattern +−+− . . . . (This is pattern *a* with elements 4 through 7 cleared.) Describe the time course and build-up of activation of the units in terms of the similarity of the input pattern to the two stored patterns.

Q.6.2.3. Present the pattern +−+−++−+ using the *test* command. This pattern is identical to pattern *a* except for the last element. Describe the time course and build-up of activation of the units. Specifically explain the activations of units 6 and 7 at the end of cycles 1, 2, 3, 4, 5, and 6.

Prototype learning and self-connections. In this exercise, we will see how the BSB model can be used to learn the prototype of a set of training experiences. For this purpose, reset the weights in the network and read in the patterns in the file *bsb8.pat*. This file contains eight patterns, each made by changing a different bit in the pattern +−+−+−+−. Set *ncycles* back to 2, and train the network for one epoch using these patterns. Note that during this training, the network sees eight different distortions of the prototype and never sees the prototype itself. After training, set *ncycles* back to 8 before testing.

Q.6.2.4. Test the network with one of the eight training patterns and with the prototype pattern (using the *test* command). Describe what happens in each case and explain, referring to the values of the weights. Explain the values of the weights in terms of the correlational character of Hebbian learning.

Ex. 6.3. The Delta Rule in a Linear Associator

This exercise allows you to explore the delta rule in a linear auto-associator. First you will get a chance to develop your own set of training patterns that are not orthogonal yet still "solvable" in an auto-associative network. Then you will get a chance to try to break the network by finding a set of patterns that the network cannot solve.

When learning using the delta rule, there is always a trivial solution to any auto-associator problem as long as there are self-connections of the units: The trivial solution is to make the self-connections large enough so that each unit essentially generates its own internal input to match the external input it receives from the outside. In many cases (as you can demonstrate in the exercises) the network will in fact make use of this situation to set the self-connections to 1.0. For this reason, the delta rule is typically used without self-connections in an auto-associator.

Begin this exercise by constructing a set of three nonorthogonal patterns that are learnable by the network. One possibility is to generate such a set at random, using the *get/ rpatterns* command. Of course, if you do this you will have to check to make sure that they are learnable. Alternatively, you can try constructing a set on your own.

For this exercise you will begin by calling the program with the *dr8.tem* and *.str* files:

aa dr8.tem dr8.str

The template file differs from the one you have been using up to now only in that it displays the *pss* (pattern sum of squares) and *tss* (total sum of squares) measures. The *dr8.str* file sets *linear* mode to 1, but all other modes have their default values; in particular, *hebb* mode and *selfconnect* mode are both off. Since training takes several epochs to run to completion, the file also sets *stepsize* to *pattern* and sets *nepochs* to 10. This means that the display is updated only once for each pattern after cycling and changing the weights. The *ncycles* variable is set to 2. The activation at the end of processing reflects the sum of the external input and the internal input generated by the activations produced by external input on cycle 1. When the internal input matches the external input, the error term is 0. Thus, when then network has "solved" a set of training patterns, the activations will equal twice the external input. Finally, the *lrate* parameter is set to 0.05; large values result in overshooting weight changes, which can lead to disaster.

The linear predictability constraint. The first exercise is to generate your set of three nonorthogonal but still learnable patterns, and present them to the network for learning. Study what happens to the *pss* and *tss* measures (you may want to set *single* to 1 to do this). Run the learning process to the point where the *tss* becomes less than 0.001 (*strain* terminates when this occurs since *ecrit* is set to 0.001), and compare the resulting weights to those that you obtain if you teach the network the same set of patterns for 10 epochs using the Hebb rule.

Q.6.3.1. Display your three patterns, and discuss the two weight matrices you obtained. In what ways is the matrix obtained using the delta rule similar to the matrix obtained using Hebbian learning? How are they different? Explain.

Hints. We supply a set of patterns in *dr8.pat*, which you may use if you wish. One of the things you will observe in the delta rule solution is that the weights on each row of the weight matrix add up to 0.99; when the problem is completely solved, and if you could see all the decimal places, they would add up to 1.0. This is not the case in the Hebbian matrix. Your answer should include an explanation of this fact.

An unsolvable set of patterns. This exercise is a little more difficult. The task is to construct a set of patterns such that there is no set of weights that will reduce the *tss* to 0. You will want to test your set of patterns, of course, to be sure that it cannot be solved. Run *strain* until it is clear that the *tss* is not getting any better. We supply a file called *imposs.pat* containing a set of patterns that cannot be learned, but it is better to make up your own.

Q.6.3.2. Display your set of unsolvable patterns, and explain how you arrived at this set. Then show the set of weights obtained by the network and indicate where the problem lies in solving the set of patterns.

Hints. It is possible to construct unsolvable sets of only two patterns, in which the two patterns are identical except for a single unit; this unit cannot then be predicted from the other units. These cases are somewhat trivial, since they would be unsolvable by any network. More interesting are the cases that involve, say, four patterns. Here, sets can be constructed that could be solved by a learning mechanism that can make use of hidden units, but not by a one-layer associative network. For simplicity, it is worthwhile to focus on constructing a set of patterns in which only one of the units is "unsolvable." It may be necessary to set the *lrate* to 0.01 to avoid wild oscillations of the weights.

Q.6.3.3. What happens if you set up the network so that two units are perfectly correlated with each other but neither can be predicted by any of the other units? Discuss the implications of this for using the auto-associator as a pattern completion device.

Ex. 6.4. Memory for General and Specific Information

Our final exercise with the **aa** program allows you to explore the version of the auto-associator discussed in *PDP:17* and *PDP:25*—the DMA model. You should be able to use the program to set up and replicate all of the simulation experiments described in *PDP:17* and *PDP:25*. We have provided files to allow you to repeat variants of several of the examples discussed in *PDP:17* (pp. 182-192). These examples illustrated the following aspects of the DMA model:

1. The model can extract what appears to be the prototype, or central tendency, of a set of patterns if the patterns are in fact random distortions of the same base or prototype pattern.

2. The model can do this for several different patterns, using the same set of connections to store its knowledge of all of the prototypes.

3. This ability does not depend on the exemplars being presented with labels.

4. Representations of specific repeated exemplars can coexist in the same set of connections with knowledge of the prototype.

The examples discussed in *PDP:17* were formulated on a 24-unit auto-associator. Patterns presented to the network consisted of an eight-unit name field and a 16-unit visual pattern field. Because the weights for a 24-unit associator cannot be displayed conveniently on the screen, the versions of these examples considered here use a 16-unit associator that implements only the visual pattern field from the *PDP:17* examples.

Learning a prototype from exemplars. We have already explored this capability in Ex. 6.2, using a systematic set of distortions. In this instance each distortion is a random variant of the prototype. A single prototype pattern—taken, for concreteness, to represent the typical dog—is distorted randomly on each learning trial by changing the sign of each element of the external input with a probability given by the value of the parameter *pflip*. The weights are adjusted after each pattern according to the delta rule. In *PDP:17*, we imagined that the learning rate parameter was rather large, so that immediately after the weight adjustment the weights reflected the new exemplar, but that each new increment decayed to a fraction of its initial value before the next pattern was presented. Here we simply set the learning rate parameter to a small value so that the weight increments already reflect the assumed decay.

The parameter values used in this simulation are as follows: *estr* = 0.15, *istr* = 0.15, and *decay* = 0.15. The learning rate parameter *lrate* and the distortion rate parameter *pflip* are set initially to 0.01[1] and 0.2, respectively,

[1] Careful reading of the text of *PDP:17* suggests that the example described in the section "Learning a Prototype From Exemplars" would have used a learning rate of (0.05)(0.85)/(*nunits* − 1) which is about 0.002. This is not correct; a value of 0.01 reproduces the results described more closely, so we have used the latter value here. The difference is one of the magnitude of the impact of each distortion to the weights and does not affect the qualitative character of the results.

but the experiments we suggest involve manipulating these parameters. Also, the number of processing cycles run in processing each pattern (*ncycles*) is set to 10 for these examples. Larger values will allow the network to settle closer to asymptotic activation values but do not affect the results materially so we use a smaller value to save computing time.

The example is carried out using the *dma.tem* and *dma.str* files to set up the screen layout and set the parameters to the appropriate values. Then the file *dog.pat* is read in using the *get/ patterns* command. The screen layout is rather cramped so that all of the information displayed previously still fits. To train the network with a series of distorted exemplars to the *dog* prototype, simply enter the *strain* command. You will want to enter it twice to complete 50 epochs. The prototype and an approximate characterization of the weights that should result are shown in Figure 3. Note that the

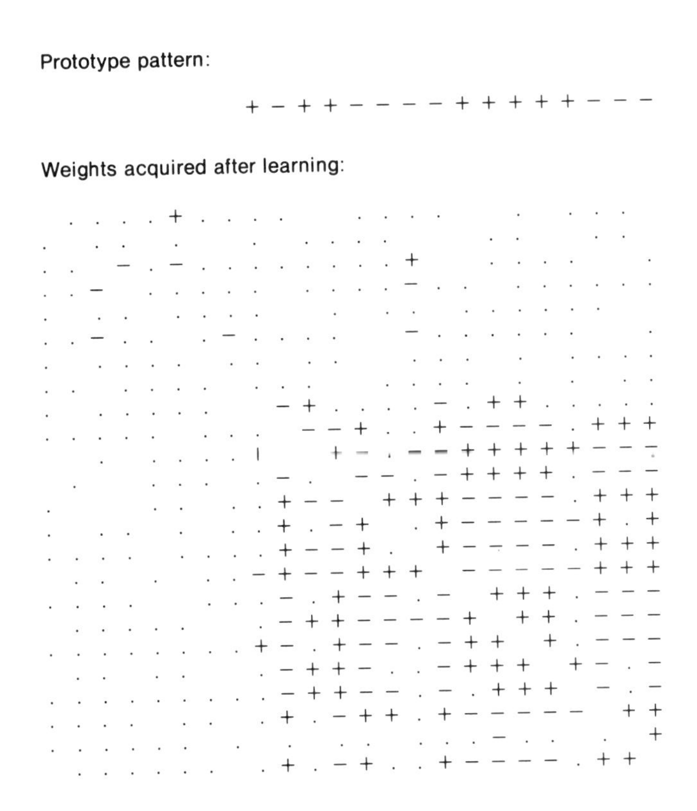

FIGURE 3. Weights acquired in learning from distorted exemplars of a prototype. (The prototype pattern is shown above the weight matrix. Blank entries correspond to weights with absolute values less than 0.01; dots correspond to absolute values less than 0.06; pluses or minuses are used for weights with larger absolute values.) (From "Distributed Memory and the Representation of General and Specific Information" by J. L. McClelland and D. E. Rumelhart, 1985, *Journal of Experimental Psychology*, *114*, 159-188. Copyright 1985 by the American Psychological Association. Reprinted by permission.)

display you will see on your screen corresponds to the lower left 16×16 entries in this figure.

At the end of 50 epochs, test the network using the *test* command. Test it on the last exemplar studied (using the *L* option with *test*), on the prototype (by entering *#dog* to the prompt), and on two or three new distortions (by entering *?dog* to the prompt). Pay particular attention to the *ndp* measure. Then set *nepochs* to 1 and run several more individual epochs of training, testing the model as just described after each epoch.

Q.6.4.1. Compare the weights you obtain to the results shown in Figure 3. Are they qualitatively similar? Also describe the results of the tests. Discuss the sense in which this model is sensitive both to the prototype and to recent, specific exemplars.

Q.6.4.2. Repeat the experiment using larger and smaller values of *lrate* and larger and smaller values of *pflip*. Describe the results and explain. Note that the network will repeat the exact same series of distortions on successive runs in which *pflip* is not changed if reinitialization is done with *reset*; a new series of distortions will occur if *newstart* is used to reinitialize. If *pflip* is decreased (or increased) and *reset* is used, the distortions should be a subset (or superset, respectively) of the distortions from the previous run.

Learning several categories without labels. This example allows you to replicate the dog-cat-bagel example discussed on pages 184-189 of *PDP:17*. Once again the name field will not be used, so that the experiments will be most comparable to those discussed in the section "Category Learning Without Labels" in *PDP:17*.

In the dog-cat-bagel example, there are three prototypes: the same one you have been using as the *dog* and two new ones—the *cat*, similar to *dog*, and the *bagel*, orthogonal to both of these. The network is trained using distortions of each of these, with *pflip* set to 0.1. The prototypes and the weights that resulted from training with both name and visual patterns are shown in Figure 4, and the results of learning the three prototypes without names are reproduced in Table 1.

To run the exercise, increase *nepochs* and set the *lrate* and *pflip* parameters to 0.01 and 0.1, respectively. Use *newstart* to reinitialize the network, and read in the three patterns from the file *dcb.pat* using the *get/ patterns* command. Then run 50 training epochs.

Q.6.4.3. Compare the weights you obtain to those shown in the lower right portion of Figure 4. Are there any systematic differences? Why might such differences occur?

Q.6.4.4. Use the *test* command to enter the probes shown in Table 1, and compare the pattern of activation you obtain to that shown in the

Prototypes

Dog:

```
+ - + - + - + -     + - + + - - - - + + + + + - - -
```

Cat:

```
+ + - - + + - -     + - + + - - - - + - + - + + - +
```

Bagel:

```
+ - - + + - - +     + + - + - + + - + - - + + + + -
```

Weights

```
 -1 -1   +2*-1   -1     +3       +5 -2 -1    -3 +2 -1       +2       -1
      -2   -1 +4*       +1 -1           -1 -1 +1 -1 -1    -3 -1   -3 +3
-1 -2    -1       +5          +1    +1 -1 -1    -1 +4   +1    -5    -1
      -1    -1    -1 +2    -1 +3 -3       +2 +3        -1 -2    -1 +1 +3
+3 -1 -1              +1     -1    +3 -3       -2 +2        +1 +3 +1 -1 -1
-1 +5 -3            -2     -1 -1    -1     -1          -1 -2 +1 -5 -1 +2 -1 +5
-1    +3 -1 -1 -2                 +1        -1    +1     +4       -1 -2
-1 -1 -1 +2          -1    -1 +3 -2 -1     +3 +3    -1 -1 -3 +1 -1 +3 +3

+3    -1 -1 +2 -1 -1                 +3 -2        -3 +2    +1     +3 +1    -1
-1 -1 -1 +3    -1    +3           -3 -1     +3 +3 +1 -1 -2 -2 +1 -1     +2
+1 +1 +1 -3            -2    +1 -3        -1 -3 -3         +1 +3 -1    -1 -2 +2
+2 -1 -1 -1 +3    -1          +3 -1 +1    -2 -1      -2 +2 -1          +2 +1 -1
-2 +1 +1     -2 +1 +1 +1     -3          -3          +1 +2 -3         -1 -2 -1 +1
    1     |2   1          +2    -1 +2 -2     +1     +? +1 -1      -3     -1     +2
-1 -1 -1 +2     -1 -1 +2         +2 -3          +3     +1          -3     -1     +3
-2 +1 +1 +1 -3      +1 +1    -2 +1 -1 -3 +3              -2 +1 -1 -1 -3 -1 +1
+3    -1     +2     -1       +2          +3 -3          -3     -1 +1     +3 +1 -1 -1
      +5 -1 -1 -1 +4 -1                       +1     -3 +1         +1     -1 -5 -1 -1
   +1     -2     +1 +1 -3    +1 -2 +2 +1 -1 -3 -3 -1    +1     -2          -3 +1
+1 -5 +2 +1 +1 -5 +3                 -2 +1 -1 +1          +1 +1 -2     +1 -2 +1 -3
+2 -1 -1     +3 -1           +2     +1 +2 -2           -3 +2     +1
   +3 -4 +1      +3 -4 +1    +1     -1          +1 +1     +1 -5 -1 -3          +1 +2
         +2 -1      -1       -1 +2 -3          +2 +2          -1 -3          +1
-1 +3 -2 -1 -1 +5 -3             -1 +3     +1 -1          -1     +1 -2     +3 -1
```

FIGURE 4. Weights acquired in learning the three prototype patterns shown. (Blanks in the matrix of weights correspond to weights with absolute values less than or equal to 0.05. Otherwise the actual value of the weight is about 0.05 times the value shown; thus +5 stands for a weight of +0.25. The gap in the horizontal and vertical dimensions is used to separate the name field from the visual pattern field.) (From "Distributed Memory and the Representation of General and Specific Information" by J. L. McClelland and D. E. Rumelhart, 1985, *Journal of Experimental Psychology*, *114*, 159-188. Copyright 1985 by the American Psychological Association. Reprinted by permission.)

TABLE 1

RESULTS OF TESTS AFTER LEARNING THE
DOG, CAT, AND BAGEL PATTERNS WITHOUT NAMES

Dog visual pattern:	+	−	+	+	−	−	−	−	+	+	+	+	+	−	−	−
Probe:									+	+	+	+				
Response:	+3	−3	+3	+3	−3	−4	−3	−3	+6	+5	+6	+5	+3	−2	−3	−2
Cat visual pattern:	+	−	+	+	−	−	−	−	+	−	+	−	+	+	−	+
Probe:									+	−	+	−				
Response:	+3	−3	+3	+3	−3	−3	−3	−3	+6	−5	+6	−5	+3	+2	−3	+2
Bagel visual pattern:	+	+	−	+	−	+	+	−	+	−	−	+	+	+	+	−
Probe:									+	−	−	+				
Response:	+2	+3	−4	+3	−3	+3	+3	−3	+6	−6	−6	+6	+3	+3	+3	−3

Note. From "Distributed Memory and the Representation of General and Specific Information" by J. L. McClelland and D. E. Rumelhart, 1985, *Journal of Experimental Psychology*, *114*, 159-188.

table. What happens when you increase *ncycles* to 50? Repeat the experiment, using other portions of the patterns as probes. How well do you feel the model does in completion of the prototypes compared to what you might expect from human subjects?

Hints. To do these tests, you must enter *E* to the prompt presented by the *test* routine, and then enter the pattern you wish to test, with dot (.) or 0 in each location that should be blank in the probe. The entries should be separated by spaces and terminated with the word *end* or an extra *return*.

Coexistence of the prototype and repeated exemplars. Our last experiment involves examining the model's ability to retain both prototypes and frequently recurring exemplars. The example is modeled on the one described in *PDP:17* on pages 189-191. In that example, we assumed the model sees several distortions of the same prototype, intermixed with presentations of two repeated examples. The patterns came with names in this example; each new distortion of the prototype was just called *dog*, but the repeated examples were called *Fido* and *Rover*.

For this example, we imagine that the 16-unit network consists of an eight-unit name field and an eight-unit visual pattern field, and we use the name pattern and the first and last groups of four elements from the visual patterns used in *PDP:17*. The original patterns are shown in Table 2. The portion of the visual pattern that is used in the present example is underlined. Note that the elements that differ in the original between the

TABLE 2

RESULTS OF TESTS WITH PROTOTYPE AND SPECIFIC EXEMPLAR PATTERNS

	Name Pattern	Visual Pattern
Pattern for dog prototype	+ − + − − − + −	+ − + + − − − − + + + + + − − −
Response to prototype name		+4 −5 +3 +3 −4 −3 −3 −3 +3 +3 +4 +3 +4 −3 −4 −4
Response to prototype visual pattern	+5 −4 +4 −4 +5 −4 +4 −4	
Pattern for Fido exemplar	+ − − − − − − −	+ − (−) + − − − − + + + + + (+) − −
Response to Fido name		+4 −4 −4 +4 −4 −4 −4 −4 +4 +4 +4 +4 +4 +4 −4 −4
Response to Fido visual pattern	+5 −5 −3 −5 −4 −5 −3 −5	
Pattern for Rover exemplar	+ − − + + + − +	+ (+) + + − − − − + + + + + − − −
Response to Rover name		+4 +5 +4 +4 −4 −4 −4 −4 +4 +4 +4 +4 +4 −4 −4 −4
Response to Rover visual pattern	+4 −4 −2 +4 −4 +4 −2 +4	

Note. Underlined portions of vectors are those used in the patterns stored in the file *dfr.pat*. Elements in parentheses are those that distinguish the visual pattern for each repeated exemplar (Fido and Rover) from the prototype. (Adapted from "Distributed Memory and the Representation of General and Specific Information" by J. L. McClelland and D. E. Rumelhart, 1985, *Journal of Experimental Psychology*, *114*, 159-188. Copyright 1985 by the American Psychological Association. Reprinted by permission.)

prototype and the two exemplars are retained in the shorter versions used here. In the version of this example we consider here, distortions apply to both parts of the pattern (name and visual parts) and occur on the repeated exemplars, as well as the prototype itself. This changes the results in some details, mainly increasing the number of epochs required for adequate learning, but the basic qualitative results are the same as reported in *PDP:17*.

This example is run with the same parameters as the previous one, so all you have to do is reinitialize the network and read in the *dog*, *Fido*, and *Rover* patterns from the file *dfr.pat*. (If you start again from scratch, do not forget to set *pflip* to 0.1.) Use *strain* (or *ptrain*, if you prefer) to run 50 epochs of training, and then test the network using the *ctest* routine, testing first for the network's ability to fill in the visual pattern from the name and then testing the network's ability to fill in the name pattern from the visual pattern. (Set *ncycles* to 50 for these tests.) These tests require clearing elements 8 through 15 and 0 through 7, respectively. For example, to test for the network's ability to fill in the visual pattern for *dog*, you would enter:

ctest dog 8 15

Set *ncycles* back to 10, run another 50 epochs, set *ncycles* to 50 again, and test again.

Q.6.4.5. How close do you come to reproducing the results shown in Table 2 at the end of 50 epochs? At the end of 100 epochs? In general, what do you see as the strengths and weaknesses of this approach to storing and retrieving representations of categories and exemplars?

COMPETITIVE LEARNING

In Chapter 5 we showed that multilayer, nonlinear networks are essential for the solution of many problems. We showed one way, the back propagation of error, that a system can learn appropriate features for the solution of these difficult problems. This represents the basic strategy of pattern association—to search out a representation that will allow the computation of a specified function. There is a second way to find useful internal features: through the use of a *regularity detector,* a device that discovers useful features based on the stimulus ensemble and some a priori notion of what is important. The competitive learning mechanism described in *PDP:5* is one such regularity detector. In this section we describe the basic concept of competitive learning, show how it is implemented in the **cl** program, describe the basic operations of the program, and give a few exercises designed to familiarize the reader with these ideas.

BACKGROUND

The basic architecture of a competitive learning system (illustrated in Figure 5) is a common one. It consists of a set of hierarchically layered units in which each layer connects, via excitatory connections, with the layer immediately above it, and has inhibitory connections to units in its own layer. In the most general case, each unit in a layer receives an input from each unit in the layer immediately below it and projects to each unit in the layer immediately above it. Moreover, within a layer, the units are broken into a set of inhibitory clusters in which all elements within a cluster inhibit all other elements in the cluster. Thus the elements within a cluster at one level compete with one another to respond to the pattern appearing on the layer below. The more strongly any particular unit responds to an incoming stimulus, the more it shuts down the other members of its cluster.

There are many variants to the basic competitive learning model. Von der Malsburg (1973), Fukushima (1975), and Grossberg (1976), among others, have developed competitive learning models. In this section we describe the simplest of the many variations. The version we describe was first proposed by Grossberg (1976) and is the one studied by Rumelhart and Zipser in *PDP:5*. This version of competitive learning has the following properties:

- The units in a given layer are broken into several sets of nonoverlapping clusters. Each unit within a cluster inhibits every other unit within a cluster. Within each cluster, the unit receiving the largest input achieves its maximum value while all other units in the cluster are pushed to their minimum value.[2] We have arbitrarily set the maximum value to 1 and the minimum value to 0.

- Every unit in every cluster receives inputs from all members of the same set of input units.

- A unit learns if and only if it wins the competition with other units in its cluster.

[2] A simple circuit, employed by Grossberg (1976) for achieving this result, is attained by having each unit activate itself and inhibit its neighbors. Such a network can readily be employed to *choose* the maximum value of a set of units. In our simulations, we do not use this mechanism. We simply compute the maximum value directly.

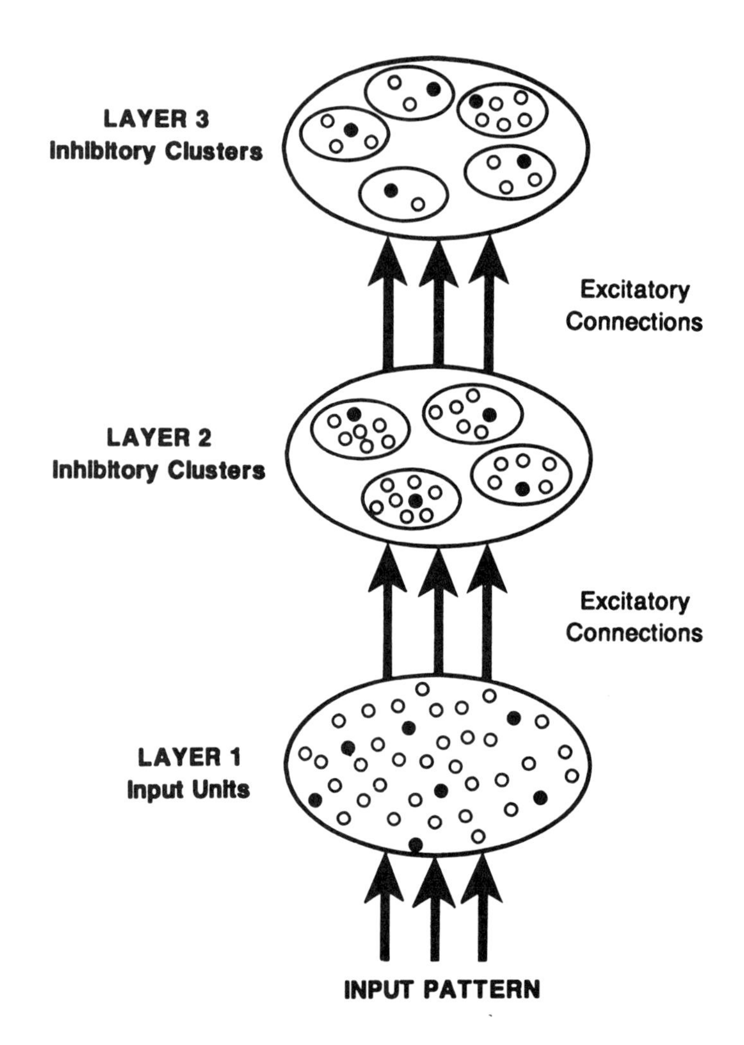

FIGURE 5. The architecture of the competitive learning mechanism. Competitive learning takes place in a context of sets of hierarchically layered units. Units are represented in the diagram as dots. Units may be active or inactive. Active units are represented by filled dots, inactive ones by open dots. In general, a unit in a given layer can receive inputs from all of the units in the next lower layer and can project outputs to all of the units in the next higher layer. Connections between layers are excitatory and connections within layers are inhibitory. Each layer consists of a set of clusters of mutually inhibitory units. The units within a cluster inhibit one another in such a way that only one unit per cluster may be active. We think of the configuration of active units on any given layer as representing the input pattern for the next higher level. There can be an arbitrary number of such layers. A given cluster contains a fixed number of units, but different clusters can have different numbers of units. (From "Feature Discovery by Competitive Learning" by D. E. Rumelhart and D. Zipser, 1985, *Cognitive Science*, 9, 75-112. Copyright 1985 by Ablex Publishing. Reprinted by permission.)

- A stimulus pattern S_j consists of a binary pattern in which each element of the pattern is either *active* or *inactive*. An active element is assigned the value 1 and an inactive element is assigned the value 0.

- Each unit has a fixed amount of weight (all weights are positive) that is distributed among its input lines. The weight on the line connecting to unit i on the upper layer from unit j on the lower layer is designated w_{ij}. The fixed total amount of weight for unit j is designated $\sum_j w_{ij} = 1$. A unit learns by shifting weight from its inactive to its active input lines. If a unit does not respond to a particular pattern, no learning takes place in that unit. If a unit wins the competition, then each of its input lines gives up some portion ϵ of its weight and that weight is then distributed equally among the active input lines. Mathematically, this learning rule can be stated

$$\Delta w_{ij} = \begin{cases} 0 & \text{if unit } i \text{ loses on stimulus } k \\ \epsilon \dfrac{active_{jk}}{nactive_k} - \epsilon w_{ij} & \text{if unit } i \text{ wins on stimulus } k \end{cases}$$

where $active_{jk}$ is equal to 1 if in stimulus pattern S_k, unit j in the lower layer is active and is zero otherwise, and $nactive_k$ is the number of active units in pattern S_k (thus $nactive_k = \sum_j active_{jk}$).[3]

Figure 6 illustrates a useful geometric analogy to this system. We can consider each stimulus pattern as a vector. If all patterns contain the same number of active lines, then all vectors are the same length and each can be viewed as a point on an N-dimensional hypersphere, where N is the number of units in the lower level, and therefore, also the number of input lines received by each unit in the upper level. Each × in Figure 6A represents a particular pattern. Those patterns that are very similar are near one another on the sphere, and those that are very different are far from one another on the sphere. Note that since there are N input lines to each unit in the upper layer, its weights can also be considered a vector in N-dimensional space. Since all units have the same total quantity of weight, we have N-dimensional vectors of approximately fixed length for each unit

[3] Note that for consistency with the other chapters in this book we have adopted terminology here that is different from that used in the *PDP:5*. Here we use ϵ where g was used in *PDP:5*. Also, here the weight to unit i from unit j is designated w_{ij}. In *PDP:5*, i indexed the sender not the receiver, so w_{ij} referred to the weight from unit i to unit j.

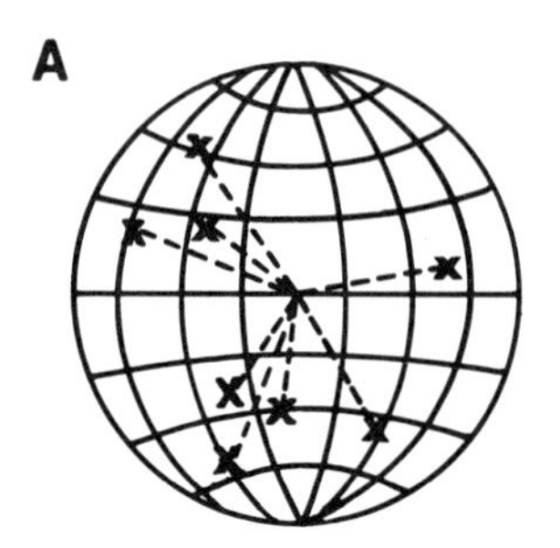

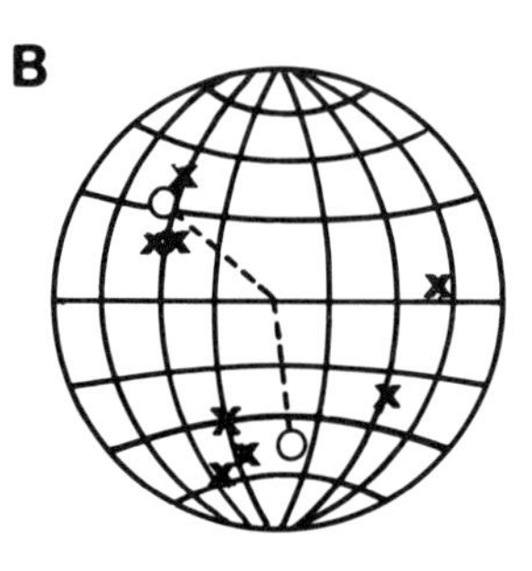

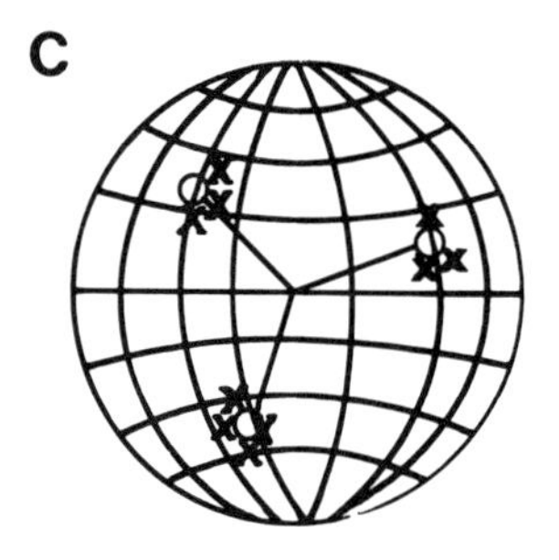

FIGURE 6. A geometric interpretation of competitive learning. *A:* It is useful to conceptualize stimulus patterns as vectors whose tips all lie on the surface of a hypersphere. We can then directly see the similarity among stimulus patterns as distance between the points on the sphere. In the figure, a stimulus pattern is represented as an $\times$. The figure represents a population of eight stimulus patterns. There are two clusters of three patterns and two stimulus patterns that are rather distinct from the others. *B:* It is also useful to represent the weights of units as vectors falling on the surface of the same hypersphere. Weight vectors are represented in the figure as ○'s. The figure illustrates the weights of two units falling on rather different parts of the sphere. The response rule of this model is equivalent to the rule that whenever a stimulus pattern is presented, the unit whose weight vector is closest to that stimulus pattern on the sphere wins the competition. In the figure, one unit would respond to the cluster in the northern hemisphere and the other unit would respond to the rest of the stimulus patterns. *C:* The learning rule of this model is roughly equivalent to the rule that whenever a unit wins the competition (i.e., is closest to the stimulus pattern), that weight vector is moved toward the presented stimulus. The figure shows a case in which there are three units in the cluster and three natural groupings of the stimulus patterns. In this case, the weight vectors for the three units will each migrate toward one of the stimulus groups. (From "Feature Discovery by Competitive Learning" by D. E. Rumelhart and D. Zipser, 1985, *Cognitive Science*, 9, 75-112. Copyright 1985 by Ablex Publishing. Reprinted by permission.)

in the cluster.[4] Thus, properly scaled, the weights themselves form a set of vectors that (approximately) fall on the surface of the same hypersphere. In Figure 6B, the o's represent the weights of two units superimposed on the same sphere with the stimulus patterns. Whenever a stimulus pattern is presented, the unit that responds most strongly is simply the one whose weight vector is nearest that for the stimulus. The learning rule specifies that whenever a unit wins a competition for a stimulus pattern, it moves a fraction ϵ of the way from its current location toward the location of the stimulus pattern on the hypersphere. Suppose that the input patterns fell into some number, M, of "natural" groupings. Further, suppose that an inhibitory cluster receiving inputs from these stimuli contained exactly M units (as in Figure 6C). After sufficient training, and assuming that the stimulus groupings are sufficiently distinct, we expect to find one of the vectors for the M units placed roughly in the center of each of the stimulus groupings. In this case, the units have come to detect the grouping to which the input patterns belong. In this sense, they have "discovered" the structure of the input pattern sets.

Some Features of Competitive Learning

There are several characteristics of a competitive learning mechanism that make it an interesting candidate for study, for example:

- Each cluster classifies the stimulus set into M groups, one for each unit in the cluster. Each of the units captures roughly an equal number of stimulus patterns. It is possible to consider a cluster as forming an M-valued feature in which every stimulus pattern is classified as having exactly one of the M possible values of this feature. Thus, a cluster containing two units acts as a binary feature detector. One element of the cluster responds when a particular feature is present in the stimulus pattern, otherwise the other element responds.

- If there is structure in the stimulus patterns, the units will break up the patterns along structurally relevant lines. Roughly speaking, this means that the system will find clusters if they are there.

[4] It should be noted that this geometric interpretation is only approximate. We have used the constraint that $\sum_j w_{ij} = 1$ rather than the constraint that $\sum_j w_{ij}^2 = 1$. This latter constraint would ensure that all vectors are in fact the same length. Our assumption only assures that they will be approximately the same length.

- If the stimuli are highly structured, the classifications are highly stable. If the stimuli are less well structured, the classifications are more variable, and a given stimulus pattern will be responded to first by one and then by another member of the cluster. In our experiments, we started the weight vectors in random directions and presented the stimuli randomly. In this case, there is rapid movement as the system reaches a relatively stable configuration (such as one with a unit roughly in the center of each cluster of stimulus patterns). These configurations can be more or less stable. For example, if the stimulus points do not actually fall into nice clusters, then the configurations will be relatively unstable and the presentation of each stimulus will modify the pattern of responding so that the system will undergo continual evolution. On the other hand, if the stimulus patterns fall rather nicely into clusters, then the system will become very stable in the sense that the same units will always respond to the same stimuli.[5]

- The particular grouping done by a particular cluster depends on the starting value of the weights and the sequence of stimulus patterns actually presented. A large number of clusters, each receiving inputs from the same input lines can, in general, classify the inputs into a large number of different groupings or, alternatively, discover a variety of independent features present in the stimulus population. This can provide a kind of distributed representation of the stimulus patterns.

- To a first approximation, the system develops clusters that minimize within-cluster distance, maximize between-cluster distance, and balance the number of patterns captured by each cluster. In general, tradeoffs must be made among these various forces and the system selects one of these tradeoffs.

IMPLEMENTATION

The competitive learning model is implemented in the **cl** program. The model implements a single input (or lower level) layer of units, each connected to all members of a single output (or upper level) layer of units. The basic strategy for the **cl** program is the same as for **bp** and the other learning programs. Learning occurs as follows: A pattern is chosen and the

[5] Grossberg (1976) has addressed this problem in his very similar system. He has proved that if the patterns are sufficiently sparse and/or when there are enough units in the cluster, then a system such as this will find a perfectly stable classification. He also points out that when these conditions do not hold, the classification can be unstable. Most of our work is with cases in which there is no perfectly stable classification and the number of patterns is much larger than the number of units in the inhibitory clusters.

pattern of activation specified by the input pattern is clamped on the input units. Next, the net input into each of the output units is computed. The output unit with the largest input is determined to be the winner and its activation value is set to 1. All other units have their activation values set to 0. The routine that carries out this computation is

```
compute_output() {

/* initialize all output units */
  for (i = ninputs; i < nunits; i++) {
    netinput[i] = 0.0;
    activation[i] = 0.0;
  }

/* compute the netinput for each output unit i */
  for (j = 0; j < ninputs; j++) {
    if (activation[j]) {
      for (i = ninputs; i < nunits; i++) {
        netinput[i] += weight[i][j];
      }
    }
  }

/* find the winner */
  for (winner = ninputs, i = ninputs; i < nunits; i++) {
    if (netinput[winner] < netinput[i]) {
      winner = i;
    }
  }

/* set the winner's activation to 1.0 */
  activation[winner] = 1.0;
}
```

After the activation values are determined for each of the output units, the weights must be adjusted according to the learning rule. This involves increasing the weights from the active input lines to the winner and decreasing the weights from the inactive lines to the winner in such a way that the total amount of weight is kept equal to 1.0. This is done by the following routine:

```
change_weights()
{

/* first we determine how many input lines are on */
  for (j = 0; j < ninputs; j++) {
    if (activation[j])
      nactive += 1;
  }
```

```
/* if no input lines are on no learning takes place */
  if(nactive == 0) return;

/* otherwise, we adjust the winner's weights */
  for (j = 0; j < ninputs; j++) {
    weight[winner][j] +=
      lrate*(activation[j]/nactive) -
        lrate * weight[winner][j];
  }
}
```

RUNNING THE PROGRAM

The **cl** program is run in much the same way as the programs already described. In general, the program is called with a *.tem* file and a *.str* file. Because of the simplicity of the **cl** architecture (each input unit is connected to each output unit and the output units form a single inhibitory cluster), a *.net* file is not needed; instead, *ninputs* and *noutputs* are defined near the top of the *.str* file. This leads the program to create a network of *ninputs* input units connected to a cluster of *noutputs* output units. The connections are all positive and sum to 1. Generally, a *.pat* file is used to specify a list of patterns for use in training and testing.

The facilities for training and testing are the same as those used in the other learning programs. The *strain* command is used to train the network using a fixed sequential order of training in each epoch. The *ptrain* command is used to train the network using a permuted order of presentations in each epoch. Both commands run *nepochs* of training, ending when interrupted. Since there is no teacher, there is no total sum of squares or error criterion.

Commands

The commands in **cl** are a subset of those for **bp** and therefore need no further explication.

Variables

The following list mentions only those variables that are new or changed in the **cl** program.

stepsize

The default *stepsize* in **cl** is *epoch*; this means that a step consists of going through each of the input patterns, presenting the pattern, computing the activations, and determining the winner, and, if *lflag* is set, changing the weights on the winner; then, after this has been done for each of the patterns, displaying the relevant variables on the screen. Other possible values of stepsize are *pattern*, which causes updating/pausing to occur after each pattern presentation, and *nepochs*, which causes updating/pausing to occur only at the end of *nepochs*.

param/ lrate

Determines the percentage of the winner's weight that is redistributed on each learning trial.

OVERVIEW OF EXERCISES

We provide two exercises for the **cl** program. The first uses the Jets and Sharks data base to explore the basic characteristics of competitive learning. The second applies competitive learning to the difficult problem of graph partitioning. A special case of this is the *dipole* problem, considered at the end of Ex. 6.6.

Ex. 6.5. Clustering the Jets and Sharks

The Jets and Sharks data base provides a useful context for studying the clustering features of competitive learning. We have prepared the files *2jets.tem*, *2jets.str*, *jets.pat*, and a couple of *.loo* files for this example. The file *jets.pat* contains the feature specifications for the 27 gang members. (The *2* in the name *2jets.tem* indicates that the network has an output cluster of two units.) The pattern file is set up as follows: The first column contains the name of each individual. The next two tell whether the individual is a Shark or a Jet, the next three columns correspond to the age of the individual, and so on. Note that there are no inputs corresponding to name units; the name only serves as a label for the convenience of the user. To run the program type

cl 2jets.tem 2jets.str

The resulting screen display (shown in Figure 7) shows the epoch number, the name of the current pattern, the output vector, the inputs, and the weights from the input units to each of the output units. Between the inputs and the weights is a display indicating the labels of each feature.

```
cl:
disp/  exam/  get/  save/  set/  clear  do  log  newstart  ptrain  quit  reset
run  strain  tall  test


epochno        0
cpname        Art

output        0 1
                                                                  weights
         input                                             unit_1        unit_2

Gang      1  0            Je Sh                            14  2          6  7
Age       0  0  1         20 30 40                          7  0  5       8  1  8
Edu       1  0  0         JH HS co                          0  2 13       4  2  5
Mar       1  0  0         si ma di                          1 14  2       9 11  6
Job       1  0  0         pu bg bo                         16  8 10      12  4 11
```

FIGURE 7. Initial screen display for the **cl** program running the Jets and Sharks example with two output units.

The inputs and weights are configured in a manner that mirrors the structure of the features. In this case, the pattern for Art is the current pattern. The first row of inputs indicate the gang to which the individual belongs. In the case of Art, we have a 1 on the left and a 0 on the right. This represents the fact that Art is a Jet and not a Shark. Note that there is at most one 1 in each row. This results from the fact that the values on the various dimensions are mutually exclusive. Art has a 1 for the third value of the *Age* row, indicating that Art is in his 40s. The rest of the values are similarly interpreted. The weights are in the same configuration as the inputs. The corresponding weight value is written below each of the two output unit labels (*unit_1* and *unit_2*). Each cell contains the weight from the corresponding input unit to that output unit. Thus the upper left-hand value for the weights is the initial weight from the *Jet* unit to output unit 1. Similarly, the lower right-hand value of the weight matrix is the initial weight from *bookie* to unit 2. The initial values of the weights are random, with the constraint that the weights for each unit sum to 1.0. (Due to scaling and roundoff, the actual values displayed should sum to a value somewhat less than 100.) The *lrate* parameter is set to 0.05. This means that on any trial 5% of the winner's weight is redistributed to the active lines. It should be noted that the *2jets.str* file has already read in the *jets.pat* pattern file.

Now try running the program using the *ptrain* command. (Note that *ptrain* is better than *strain* for the competitive learning procedure since the order can have a large effect on exactly what is learned.) Since *nepochs* is set to 20, the system will stop after 20 epochs. Look at the new values of the weights. Try several more runs, using the *newstart* command to reinitialize the system each time. In each case, note the configuration of the weights. You should find that usually one unit gets about 20% of its weight on the *jets* line and none on the *sharks* line, while the other unit shows the opposite pattern.

Q.6.5.1. What does this pattern mean in terms of the system's response to each of the separate patterns? Explain why the system usually falls into this pattern.

Hints. You can find out how the system responds to each subpattern by using the *tall* command and stepping through the set of patterns—noting each time which unit wins on that pattern (this is indicated by the output activation values displayed on the screen).

Q.6.5.2. Examine the values of the weights in the other rows of the weight matrix. Explain the pattern of weights in each row. Explain, for example, why the unit with a large value on the *Jet* input line has the largest weight for the 20s value of age, whereas the unit with a large value on the *Shark* input line has its largest weight for the 30s value of the age row.

Now repeat the problem and run it several more times until it reaches a rather different weight configuration. (This may take several tries.) You might be able to find such a state faster by reducing *lrate* to a smaller value, perhaps 0.02.

Q.5.3. Explain this configuration of weights. What principle is the system now using to classify the input patterns? Why do you suppose reducing the learning rate makes it easier to find an unusual weight pattern?

We have prepared a pattern file, called *ajets.pat*, in which we have deleted explicit information about which gang the individuals represent. Load this file by typing

get patterns ajets.pat

Q.5.4. Repeat the previous experiments using these patterns. Describe and discuss the differences and similarities.

Thus far the system has used two output units and it therefore classified the patterns into two classes. We have prepared a version with three output units. This version can be accessed by the command:

cl 3jets.tem 3jets.str

Q.6.5.5. Repeat the previous experiments using three output units. Describe and discuss differences and similarities.

Ex. 6.6. Graph Partitioning

Recall that the competitive learning mechanism with n output units has a propensity to put the stimulus patterns in n classes with the classes maximally distinct and the numbers of patterns per class approximately equal. It turns out that there is an interesting and difficult problem, the graph partitioning problem, which requires just that sort of solution. The problem, roughly stated, is this: Given a connected graph of n nodes, each of which is connected to one or more other nodes in the network, divide the graph into two parts with half of the nodes in each while minimizing the number of links that connect nodes from the two different classes. We can map this problem into a problem that competitive learning can work on in the following way. We have two output units, one for each of the two groups into which we are to classify the nodes. There must be n input units, one for each of the nodes in the graph. There is a stimulus pattern for each of the links of the graph. Each stimulus pattern consists of two units on and the rest off. If there is a link from unit i to unit j in the graph, then there is a pattern with units i and j both turned on and the rest turned off. We have prepared a very simple example of this. Files *graph.tem, graph.str, graph.pat*, and so on contain the relevant information. Figure 8 shows the initial screen layout and the graph in question. In this case, the graph

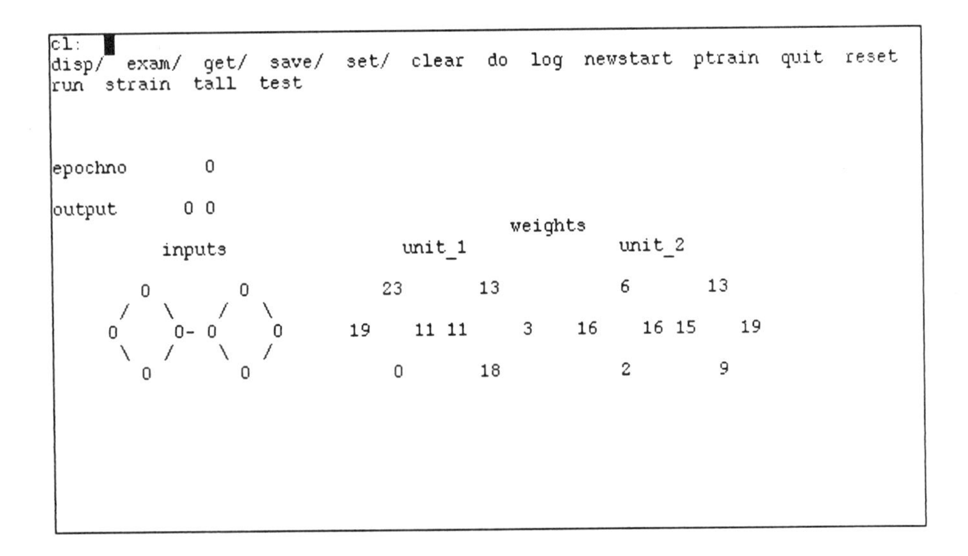

FIGURE 8. Initial screen display for the graph problem.

consists of two clusters of nodes with a single link between them. The best solution is obviously to separate the two lobes and cut the single link between. Now run the program with the command

cl graph.tem graph.str

Q.6.6.1. Using *ptrain*, run the program several times and note the solutions the system finds. To what degree do these solutions solve the graph partitioning problem? Explain the observed results. Try the same thing with various values for *lrate*. How does this change the results? Why?

Q.6.6.2. Create your own graph and evaluate the results of the network with respect to the graph partitioning problem.

It might be noted that the dipole problem, discussed at length in *PDP:5* (pp. 170-177) is an example of a rather simple graph partitioning problem. In this problem, the patterns consist of pairs of adjacent points on a two-dimensional grid (adjacent points are points that are next to each other on the same row or column). This is equivalent to the graph partioning problem in which each point is connected to every adjacent point. A set of files called *16.tem, 16.str*, and *16.pat* (together with associated *.loo* files) are provided for this example, if you choose to explore it. Just start up the **cl** program with the files *16.tem* and *16.str* and enter *ptrain* to train the network, as in all of the examples already discussed.

CHAPTER 7

Modeling Cognitive Processes: The Interactive Activation Model

In this chapter our goal is to consider the application of PDP modeling techniques to the task of accounting for human cognitive processes, as revealed through psychological experimentation. As our example for this, we've chosen the interactive activation model of word perception (McClelland & Rumelhart, 1981; Rumelhart & McClelland, 1982). This model exemplifies our approach to modeling psychological processes, and it is of tractable size for running on smaller machines.

BACKGROUND

Our initial interest in parallel distributed processing mechanisms grew out of an attempt to capture our ideas about continuous, interactive processes, particularly as they applied to the problems of visual word recognition and reading. Both of us had already done both experimental and theoretical work in this area, but without the benefit of simulations (see McClelland 1976, 1979; Rumelhart, 1977; Rumelhart & Siple, 1974).

Our primary aim was to account for contextual influences on perception. These influences have been described since psychologists first began to present visual or auditory stimuli under controlled conditions (Bagley, 1900; Cattell, 1886). Among the early observations was the fact that subjects could identify far more letters from a single brief flash if the letters fit together to form a word than if the letters made a random string. Context could override the sensory input too, as in the cases where subjects reported the strong impression that they saw all the letters in the word *FOREVER* when in fact *FOYEVER* was shown (Pillsbury, 1897).

In early studies the experimenter relied on what is generally called a free report of the contents of the briefly displayed stimulus, and many researchers pointed out that there were serious methodological problems with this. It has often been pointed out that subjects might see as much in the two cases, but forget less or correctly guess more when the stimuli form words.

An experiment that controlled for both guessing and forgetting at once was carried out by Reicher (1969) and followed up by a number of investigators, including Wheeler (1970), Johnston and McClelland (1973, 1974, 1980; Johnston, 1978; McClelland & Johnston, 1977), and several others (Baron & Thurston, 1973; Manelis, 1974; Massaro, 1973; Massaro & Klitzke, 1979; Spoehr & Smith, 1975).

In Reicher's experiments, a target was presented (e.g., *E*) either in a word (e.g., *READ*), in a scrambled letter string (e.g., *AEDR*), or in isolation. The presentation was followed by a masking stimulus, which consisted of a jumbled array of letter parts, and a pair of letters, which was keyed to the position occupied by the target letter. One of the letters was the target letter itself, and the other was another letter that fit the context (if any) to make an item of the same type. For the displays *READ, AEDR*, and *E* in isolation, the pair could be *E* and *O*, presented with a row of dashes to indicate which display location was being tested:

```
   E
 - - - -
   O
```

The subject's task was to choose which of the two letters had appeared in the indicated position. The target could appear in any of the four positions, and subjects did not know in advance which position would be tested on a given trial.

Reicher's test is called the *forced choice test*. Using this test, he found that subjects were more accurate when the letters occurred in words than in either of the other two conditions. This finding is called the *word superiority effect*.

Reicher's finding is important because it indicates that the advantage for words is not simply a matter of guessing letters that fit the context better from fragmentary cues. Rather, it appears that the perceptual system is better able to use the information in the display when the letters form a word with their context. The fact that the advantage holds for words over single letters makes it difficult to view the phenomenon as a result simply of forgetting, since a single letter surely places a very light load on memory.

Reicher's findings, backed up by a large literature of further experimental tests, seemed to us to be a very clear demonstration that context plays a role in perception. We therefore set out to model this phenomenon, basing our approach on a number of basic assumptions.

Basic Assumptions of the Interactive Activation Model

Here we describe each of the basic tenets of the interactive activation model and explain why we adopted each one.

Perception occurs in a multilevel processing system. This assumption is nearly ubiquitous, and so we will give it little discussion; surely there are separate levels of representation for visual features, for words, and for larger wholes such as sentences. For our model of the processing of individual words or strings of letters, we assumed that there are at least three levels: a visual feature level, a letter level, and a word level.

Deeper levels of processing are accessed via intermediate levels. This assumption has often seemed contentious, particularly with respect to visual word recognition. We have assumed that a letter level is interposed between the feature and the word level because words appear to be defined, not in terms of their particular visual configurations, but in terms of the sequences of letters that they contain. Thus *READ*, *read*, and *rEAD* are all recognizable as words, and letters in such stimuli are all perceived better than letters in unrelated context (e.g., the *E* in *rEAD* is perceived better than the *E* in *aEDR*; cf. Adams, 1979; McClelland, 1976). Thus it would appear that readers can use their knowledge of words to perceive sequences of letters, even if the visual configuration of the input is highly novel.

If, as we assume, access to the word level is via the letter level, then sequences of letters should be more effective as masks for words than sequences of feature bundles that do not form letters. This prediction was confirmed by Johnston and McClelland (1980).

Processing is interactive. By this we mean that processing involves the simultaneous consideration of both bottom-up input information and top-down knowledge-based constraints. Our principle reason for this belief was the well-known and ubiquitous role of contextual factors in perceptual processing already alluded to above. We take the role of word context in letter perception as one example of this kind of interactive processing. Models that captured the *outcome* of the simultaneous consideration of bottom-up and top-down information had been developed by others (particularly Morton, 1969), but we wished to embody this assumption in a dynamic processing model.

Information flow is continuous. At the time we began to consider interactive processing, the predominant view among psychologists working in perceptual information processing was that information processing occurred through a sequence of discrete steps. Each step took a certain amount of time and resulted in a discrete output. However, alternatives to this view were developed during the course of the 1970s (cf. McClelland, 1979; Norman & Bobrow, 1976; Turvey, 1973). In fact, the utility of continuous information flow was pointed out quite early in the *pandemonium* model of Selfridge (1955), an early AI model designed to account for the role of context in letter recognition. For us, the assumption of continuity seemed to be required in order to capture contextual influences in word recognition (McClelland, 1976; Rumelhart, 1977). The reason is that if the word level is to influence processing at the letter level, then the letter level must be making information available to the word level before processing at the letter level is complete.

PDP models as a way of capturing these basic assumptions. We turned to PDP models because they provided a simple and direct way of making our basic assumptions about continuous, interactive processing explicit in a computational model. By assuming a processing unit for each possible hypothesis about the input at each of the three levels of processing, by allowing each unit to be working continually, updating its own activation and sending activation to other units, and by allowing units to influence each other via simple excitatory and inhibitory interactions, we found we were able to capture our basic assumptions in a simulation model and explore how well these assumptions could account for contextual influences in letter perception.

Central Questions

In developing the interactive activation model, there were several basic questions:

- Could we make a PDP embodiment of our basic assumptions account for the basic fact that word context facilitates letter perception, as established by Reicher and others?

- Could we account for the fact that subjects perceive letters in pronounceable nonwords (e.g., *REAT*) more accurately than letters in random or scrambled strings and, under some conditions, more accurately than single letters (Johnston & McClelland, 1973; Wheeler, 1970)?

- Could we apply the model to the large body of existing data and show that we could really account for the existing findings? The most important facts we considered were these: (a) The perceptual advantage for letters in words is shared with pronounceable nonwords; that is, letters in words and in pronounceable nonwords show a sizeable advantage over single letters or letters in random strings. (b) Within pronounceable words and nonwords, there is no consistent advantage for strings containing frequent letter clusters (e.g., *PEEP* or *TEEP*) compared to those containing much less frequent letter clusters (e.g., *POET* or *HOET*). Though apparent letter-cluster effects are found in some studies (see Rumelhart & McClelland, 1982, Experiment 9), other studies did not show these effects (McClelland & Johnston, 1977). Our hope was to account for both patterns of results, based on detailed aspects of the particular materials used in different experiments. (c) For letters in words, under the visual conditions in which Reicher's word superiority effect was obtained, there is no advantage for letters occurring in contexts that strongly constrain the identity of the letter (e.g., the *C* in *CLUE*: only three letters make words in the context *_LUE*) compared to letters in context that exert much weaker constraints (e.g., the *C* in *CAKE*; 10 letters make words in the context *_AKE*; Johnston, 1978). Again, however, such effects do occur in other studies. Our hope was to use the model to understand and account for these differences.

- Could we account for a set of new findings from our own laboratory? These findings were based on the use of a new technique for visual presentation, in which the letters in a four-letter string could be started and ended at different times. We found (as reported in Rumelhart & McClelland, 1982) that subjects perceived a particular letter better when the other letters with which it occurred were presented for a longer time. This was true both when the letter formed a word with the context and when it formed a pronounceable nonword with the context, but not when the letter was embedded in a random-letter string.

The approach that we took in trying to answer these questions was to begin by trying to develop a model that produced the basic perceptual facilitation advantage for letters in words when compared to letters in nonword strings; this turned out to be one of the hardest parts of the project. We tried a number of variants on the basic activation equation, as well as a wide range of different parameter values before we developed enough understanding for what we were doing to find a combination of assumptions that worked. Once we accomplished this, we began to consider the list of phenomena we wanted to account for, working our way more or less down the list just given. Our goal was to find a single formulation, together with a single set of parameter values, that would allow us to give a fairly close account of the findings discussed earlier.

Two further aspects of our approach are worth mentioning. First, we endeavored to keep the model as simple as possible, within the constraint that we preserved sufficient structure to capture our basic assumptions. For example, we used a highly simplified representation of the visual forms of letters, and we assumed that each letter in a visual display came rigidly channeled into one of four letter positions. Second, we did not attempt to obtain detailed quantitative fits of the model to the results of particular experiments. Rather, we attempted to come as close as we could to producing results that captured the major qualitative features of the phenomena.

We do not mean to suggest that detailed quantitative fits are not appropriate in many cases. Rather, we want to suggest that in this and many other cases, detailed quantitative fits may require very detailed and specific assumptions (for example, about the confusability of particular pairs of letters) that fall outside of the basic assumptions that are at issue. Attention to such assumptions may in certain instances interfere with the search for understanding of the basic principles that transcend such details. Under these circumstances, a simplified model may yield a more satisfying explanatory account, even when it is known to be wrong in some details, if it provides an explanation for the qualitative patterns observed in the empirical phenomena. (More discussion of this point may be found in Sejnowski's discussion of the role of computational models in *PDP:21*, pp. 387-389.)

We were also concerned with making sure we understood exactly what was going on in the model that allowed it to account for the phenomena. To this end, we spent a great deal of time studying the processing of individual examples so that the details of what was happening in the simulations would be clear. This kind of detailed study of individual items will be the focus of this chapter. In assessing the model, we supplemented this individual example approach by running large simulations with lists of items taken from the original experiments we were trying to understand. We omit these kinds of simulations here, although they played a central role in evaluating how well the model could account for the facts reported in particular experiments.

When we felt we had been reasonably successful in accounting for the phenomena that we wanted to account for, we began to consider whether there might not be additional experiments that we might do to test the principles underlying the approach. We were able to come up with one such experiment (Rumelhart & McClelland, 1982, Experiment 10). It will be described in more detail in one of the exercises.

THE INTERACTIVE ACTIVATION MODEL

In this section we describe the essential features of the interactive activation (IA) model. Some additional details may be found in McClelland and Rumelhart (1981).

Network Architecture

The model consists of units at each of three processing levels: the feature level, the letter level, and the word level (see Figure 1). At the feature level, there is a set of units that serves to detect features in each of four letter positions. Within each set, there is a unit for the *presence* of each of the line segments in the simple font used by Rumelhart and Siple (1974) (shown in Figure 2) and another unit for the *absence* of each such line segment. These units are said to be detectors for the different values (present, absent) of each of the possible line-segment features. This assumption allows the model to distinguish between not knowing whether a line segment is present (both units off) and knowing that a segment is not present (the absence unit on and the presence unit off). At the letter level, there are four sets of letter units, one for each position. Each set contains a unit for each of the 26 letters of the English alphabet. At the word level, there is a single set of detectors for each word in a list of 1179 four-letter words taken from the word list of Kucera and Francis (1967). The set

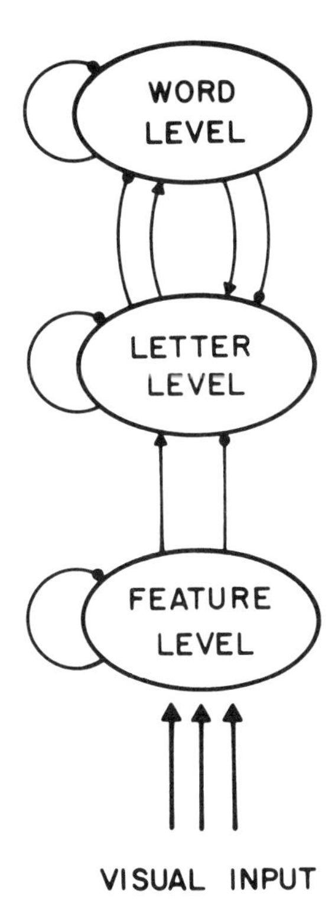

FIGURE 1. The basic architecture of the interactive activation model. (From "An Interactive Activation Model of Context Effects in Letter Perception: Part 1. An Account of Basic Findings" by J. L. McClelland and D. E. Rumelhart, 1981, *Psychological Review*, *88*, 375-407. Copyright 1981 by the American Psychological Association. Reprinted by permission.)

FIGURE 2. The Rumelhart-Siple letter font used by the interactive activation model. (From "Process of Recognizing Tachistoscopically Presented Words" by D. E. Rumelhart and P. Siple, 1974, *Psychological Review*, *81*, 99-118. Copyright 1974 by the American Psychological Association. Reprinted by permission.)

includes all the words with frequencies of two or more per million, excluding proper names, contractions, abbreviations, and foreign words that crept into the count.

Each unit can be taken as representing the hypothesis that the entity it stands for—feature, letter, or word—is present in the input. The activations of the units are monotonically related to the strengths of these hypotheses, according to a function that will be described later.

Connections

The connections among the units are intended to encode the mutual constraints among hypotheses about the possible contents of a four-letter display. The overall framework allows excitatory connections between units on different levels that are mutually consistent and allows inhibitory connections between units on different levels that are mutually inconsistent. For example, *T* in the first letter position is mutually consistent with all words beginning with *T*; these units therefore have mutually excitatory connections. To simplify matters, some of these connections are left out of the model: First, there are no feedback connections that are inhibitory. Second, there is no feedback at all from the letter level to the feature level. This leaves the following sets of between-level connections:

- *Feature-to-letter excitation.* Feature units have excitatory connections to all the letter units in the same spatial position that contain the feature. Thus, the presence unit for a horizontal bar at the top

Readout From the Model

Readout from the model is thought of as a separate process that integrates the activations of the units over time to assess the *response strength* of each unit. The response strength is defined to be the model's measure of the strength of the hypothesis that the entity a unit stands for is present in the input. The readout process chooses one of the units probabilistically, based on its response strength relative to the strength of other units.

The response strength of each unit is given by

$$s_i(t) = e^{k\bar{a}_i(t)} \tag{2}$$

where k is a scale factor and where $\bar{a}_i$, the running average of the activation of unit i, is given by

$$\bar{a}_i(t) = (orate)a_i(t) + (1 - orate)\bar{a}_i(t-1). \tag{3}$$

(Initially or when the model is reset, $\bar{a}_i(0) = a_i(0) = rest_i$, the resting activation of unit i.) Following Luce's (1959) choice model, the probability of choosing a particular item i as the response at time t is simply

$$p(r_{i,t}) = \frac{s_i(t)}{\sum_{j \epsilon C} s_j(t)}. \tag{4}$$

This probability is called the *response probability* of response i at time t. Here, C is the set of competing alternatives (all letters in the same position for letter responses; all words for word responses), including unit i itself. Note that this response choice rule has the effect of ensuring that the response probabilities always sum to 1 for each set of competing alternatives.

The Forced-Choice Test

In simulating what happens in the forced-choice test, we first made the assumption that choices were based on responses read out from the letter level only. Under this assumption, the word superiority effect is due to the feedback from the word level to the letter level.

Because the choice alternatives appear after the target display has been replaced by the mask, we assumed that readout from the network had to occur without regard to the alternatives. That is, we assumed that the subjects did their best to identify all the letters in the target display following the response choice assumptions described above, and only after doing so did they consult the forced-choice alternatives.

The determination of response probabilities in the model is complicated by the fact that the response strengths on which they are based rise and fall with time. Some assumptions must be made about the timing of readout. We assumed that subjects chose a time after onset of the target display that optimized their probability of choosing correctly. In practice, this means that readout occurs just before the onset of the postdisplay mask, if there is one. When no mask is used, readout is assumed to occur after activation and response strengths reach their asymptotic values.

Once readout has occurred, the rule for making choices is very simple: If the response letter chosen for the target position matches one of the alternatives, that alternative is chosen; otherwise, the choice is a random guess. From this rule, the probability of correctly choosing the target letter is the probability that this letter was chosen by the response readout process, plus 0.5 times the probability that neither the correct nor the incorrect alternative was chosen by the readout process.

Although the readout process is assumed to be stochastic, we do not actually simulate the probabilistic choice between alternatives. We calculate the probability of reading out the correct and incorrect alternatives and use these probabilities to calculate the probability correct in the forced choice.

Parameters

Here we discuss the parameters of the interactive activation model. There are parameters that influence the internal processing dynamics, parameters that reflect assumptions about the input, and parameters that influence the assignment of response probabilities during the readout process. The list of parameters of the model is rather long, but the majority are fixed at rather arbitrary values. The main parameters modified to capture the basic experimental effects we wished to account for are the excitatory and inhibitory strength parameters, *alpha* and *gamma*. However, in keeping with our philosophy of allowing users the greatest degree of control over the programs as possible, we have made all of the following parameters modifiable, as we shall see later.

alpha and *gamma*

The excitatory and inhibitory connection strength parameters, *alpha* and *gamma*, respectively, depend only on the processing levels of the units in question. This means, for example, that the strengths of the excitatory connections from feature units to letter units are the same for all such excitatory connections. The model has a separate parameter for feature-to-letter excitation, another for letter-to-word excitation, and another for word-to-letter excitation, as well as parameters for feature-to-letter inhibition, letter-to-letter inhibition, letter-to-word inhibition, and word-to-word inhibition. In our simulations, these parameters were subject to tuning, with the goal of obtaining the best possible fit to the entire ensemble of experimental data we were considering. The values we settled on in this process are shown in Table 1.

decay

The model provides separate parameters for the rate of decay, at both the letter and the word levels. In practice, however, the decay parameters were both set to the same value (0.07) at a fairly early point in our simulations, and then were left at this value for the remainder of our experiments with the model. This value was chosen because it seemed to be the largest value that would nevertheless allow the model to settle smoothly to a target pattern of activation. With larger values the activations of units can start to oscillate wildly from cycle to cycle.

threshold

The model provides separate parameters for the output thresholds of units at the letter and word levels. In fact, at the word level there are separate threshold values for output to the letter level and for inhibitory output to other words. These thresholds were generally left at 0, except during our simulations of the contextual enhancement effect in pronounceable nonwords (see Ex. 7.4).

max and *min*

The model also provides separate parameters for the maximum and minimum activations units are allowed to have at the letter and word levels. We have always left the maximum activations at 1.0.

TABLE 1

DEFAULT VALUES FOR THE ALPHA AND GAMMA PARAMETERS USED IN THE IA MODEL

Excitation parameters (*alpha*):	
feature to letter	0.005
letter to word	0.07
word to letter	0.30
Inhibition parameters (*gamma*):	
feature to letter	0.15
letter to word	0.04
word to word	0.21
letter to letter	0.00

We did, however, experiment with different values for the minimum, settling on −0.20 for both the letter and the word levels.

rest and *fgain*

The model provides separate parameters for the resting activation levels of units at the letter and word levels. After some experimentation, the resting levels were generally left at 0. However, in the case of words, it should be noted that the value of the word-level resting activation parameter was not 0 for all units, but was offset downward from 0 depending on the word's frequency. The most frequent word known to the model was *that*, which was assigned a resting activation offset of 0.00. Other words were given resting activation offsets ranging between −0.92 and −0.01, according to a function that assigned offsets proportional to the log of the frequency of the word, subtracted from the log of the frequency of *that*. The model multiplies these offsets by the value of a frequency-scale parameter, *fgain*, and subtracts the result from 0. Throughout the simulations we used a value of 0.05 for *fgain*. Thus, the resting levels of words actually ranged from 0.00 to nearly −0.05.

oscale

This parameter corresponds to the parameter k in Equation 2. The model provides separate parameters for scaling the output strengths of units at the letter and word levels. A larger value of *oscale* is needed at the word level to compensate for the fact that there are more competitors at this level of processing, even though most of them usually are assigned highly negative activation values by the model. We recommend that the user keep the given values of 10.0 for letter-level output and 20.0 for word-level output.

fdprob

For each display field the user wishes to present, it is possible to set a separate value for the probability that features are detected from the input. By default, *fdprob* is set to 1.0, which means that all of the features of the input are detected. However, this parameter can be set to a lower value to simulate the effects of degraded visual presentation.

estr

Many experiments find that end letters are perceived more accurately than letters internal to a word. To accommodate this, we provide separate parameters for the strength of feature-level activations for each letter position. By default, these parameters are set to 1.0, and we recommend that users leave them at these values unlcss thcy spccifically wish to explore these positional differences.

orate

This last parameter determines the rate of accumulation of activation for the purpose of determining response strength. Its default value is 0.05. Generally, we have not found the value to be terribly critical, although we have generally assumed that it is small so that activations are translated into outputs only gradually.

Processing Under Different Visual Display Conditions

In our simulations, we tended to lump visual display conditions used in different experiments into two categories, based on the different conditions used by Johnston and McClelland (1973). One condition was called the *bright-target/pattern-mask* condition and the other was called the *dim-target/blank-mask* condition.

For the bright-target/pattern-mask condition we assumed that the target display was bright and clear enough so that all visual features of the display were detected and that the same applied to the patterned masking stimulus that followed the target. We further assumed that the effect of the mask was to quickly clear out the pattern of activation at the letter level, replacing it with a new pattern.

These assumptions led us to assume that the feature-to-letter inhibition was very strong, so that features of the mask would quickly inhibit letter activations that had been produced by the target display. A side effect of this was that no letters received net bottom-up excitatory input unless they were consistent with all of the features in a particular display position. As a result, under bright-target/pattern-mask conditions, only the letter actually displayed ever became activated on the basis of bottom-up information. Therefore, we found that we had little need for letter-to-letter inhibition. Consequently, though the model provides for the possibility of such inhibitory influences, we set the letter-to-letter inhibition parameter to 0.

Given that all the features of the display are detected, why is it that performance is less than perfect in the forced-choice test? The answer is simply that it takes time for activation to build up and be read out. The role of feedback is to enhance the activation of letters and, therefore, to increase the probability of correct read out from the letter level.

For the dim-target/blank-mask conditions, we assumed that the temporal brightness summation between the target and the blank mask operated so that the display was approximately equivalent to a very low-contrast, and hence degraded, input. In this situation, we assumed that visual feature information could only be detected imperfectly. The trial is simulated as a single display of letters with a feature detection probability considerably less than 1. The effect of this is that several letters generally are consistent with the detected features in each letter position. Under these conditions, the role of feedback from the word level is to selectively enhance the activations of units at the letter level that fit together with active letters in other positions to form words or to activate groups of words.

IMPLEMENTATION

Data Structures

The implementation of the interactive activation model in the **ia** program is similar to the implementation of the IAC model, although it differs from it in many details. One important difference is that the connections among the units are not in fact specified in a connection matrix. Instead, they are determined by table look-up. There are two relevant tables: the *word* table and the *uc* table.

The *word* table, as its name implies, contains a list of all of the words known to the model, stored as a sequence of four lowercase ASCII characters. To determine whether a particular letter unit should activate a particular word, the model checks to see if the letter is in the word in the appropriate position. We do not, of course, assume that activation in the mind is actually done by table look-up.

The *uc* table contains a list of all the features of the letters. The table is called *uc* to remind the user that the model only knows one alphabet and that is the uppercase Rumelhart-Siple alphabet shown in Figure 2. Each row of the *uc* table consists of fourteen 1s and 0s, indicating whether the corresponding character does or does not have the corresponding segment from the Rumelhart-Siple font in it. For example, the row corresponding to the letter *A* is

1,1,1,1,1,0,1,0,1,0,0,0,0,0

The exact arrangement of the features will be described when we explain the use of the program. The table also contains several special, nonletter characters, in addition to the uppercase letters. These will be described later.

The *uc* table is used both to specify the set of input features, given a display specification consisting of a sequence of letters entered by the user, and to determine which letter units should be activated when a particular feature unit is activated.

One other important data structure is the *trial* data structure. This is simply a list of field onset times and their contents. This information is specified via the *trial* command, which will be described in the section on using the program.

Processing

As in the **iac** program, processing is controlled by the *cycle* routine. Here is what it looks like:

```
cycle() {

  for (cyc = 0; cyc < ncycles; cyc++) {
    cycleno++;
    if (cycleno == ftime[tt]) {
      setinput();
    }
    interact();
    wupdate();
    lupdate();
    update_out_values();
    update_display();
  }
}
```

The *setinput* routine is called when each new field is scheduled for presentation; it first zeros the activations of all of the feature units, then sets them to the new values dictated by the new input. If *fdprob* (the probability of feature detection) for the present input is less than 1, then input units are turned on only if the value of a random number returned by the random number generator is less than the value of *fdprob*. The detected input feature activations are stored in an array called *dinput*, with indexes for the feature value (absent or present), the feature (0-13), and the position (0-3).

The *interact* routine is responsible for the excitatory and inhibitory interactions between units on different levels; as we shall see, the inhibitory interactions between units on the word and letter levels are handled in the *wupdate* and *lupdate* routines. The *interact* routine has three separate parts: one for the letter-to-word interactions, one for the word-to-letter interactions, and one for the feature-to-letter interactions. For completeness we show all three portions of the routine:

```
interact() {

/* letter -> word */
  for (pos = 0; pos < WLEN; pos++) {
    for (ln = 0; ln < NLET; ln++) {
      out = l[pos][ln] - t[LU];
      if (out > 0) {
        for (wn = 0; wn < NWORD; wn++) {
          if (ln == (word[wn][pos] - 'a'))
            ew[wn] += alpha[LU] * out;
          else
            iw[wn] += gamma[LU] * out;
        }
      }
    }
  }
```

```
/* word -> letter */
  for (wn = 0; wn < NWORD; wn++) {
    out = wa[wn] - t[WD];
    if (out > 0) {
      for (ln = 0; ln < NLET; ln++) {
        for (pos = 0; pos < WLEN; pos++) {
          if (ln == (word[wn][pos] - 'a'))
            el[pos][ln] += alpha[WD] * out;
        }
      }
    }
  }

/* feature -> letter */
  for (pos = 0; pos < WLEN; pos++) {
    for (fet = 0; fet < LLEN; fet++) {
      for (val = 0; val < NFET; val++) {
        out = dinput[val][fet][pos];
        if (out > 0) {
          for (ln = 0; ln < NLET; ln++) {
            /* 0th line of table is 'A' */
            if (val == uc[ln + 'A'][fet])
              el[pos][ln] += estr[pos]*alpha[FU]*out;
            else
              il[pos][ln] += estr[pos]*gamma[FU]*out;
          }
        }
      }
    }
  }
}
```

The letter-to-word and word-to-letter portions are quite similar; we discuss just the first of these. In the letter-to-word portion, the model cycles through all the letters in each letter position. If the output of the letter unit (that is, its activation minus the letter-level threshold) is greater than 0, then the program searches through all of the words. For each word, if the word contains the letter in question in the appropriate position, then the excitation of the word is increased by the output of the unit times the letter-to-word excitation parameter; otherwise, the inhibition of the word is increased by the output of the unit times the letter-to-word inhibition parameter.

The feature-to-letter portion of *interact* is a bit different. Here, the program cycles through each of the 14 features for each letter position, first checking the absence unit (indexed by $val = 0$) for that feature, then checking the presence unit (indexed by $val = 1$). For each such unit, if it is on, the program scans through the letter table, incrementing the excitatory input to the letter units that have this feature and incrementing the inhibitory input to letter units that do not have this feature.

The Update Routines

The two update routines, *wupdate* and *lupdate*, are nearly identical; we will go through *wupdate* because it is slightly simpler—it does not have to loop separately through each of the four pools of letter-level units. The routine is as follows:

```
wupdate() {
  ss = sum;
  prsum = sum = 0;
  tally = 0;

  for (i = 0; i < NWORD; i++) {
/* word -> word inhibition */
    if (wa[i] > t[W])
      iw[i] += g[W] * (ss - (wa[i] - t[W]));
    else
      iw[i] += g[W] * ss;
/* now compute net input and update */
    net = ew[i] - iw[i];
    if (net > 0)
      effect = (max[W] - wa[i]) * net;
    else
      effect = (wa[i] - min[W]) * net;
    wa[i] += effect - decay[W]*(wa[i] - wr[i]);
    if (wa[i] > 0) {
      if (wa[i] > max[W]) wa[i] = max[W];
    }
    else {
      if (wa[i] < min[W]) wa[i] = min[W];
    }
    if (wa[i] > t[W])
      sum += wa[i] - t[W];
/* take running average for readout */
    ow[i] = ow[i] * (1 - outrate) + wa[i] * outrate;
/* then zero arrays for next cycle */
    ew[i] = iw[i] = 0;
  }
}
```

The only thing that is different about this routine compared to the *update* routine in the **iac** program is that the inhibition is handled in a way that is more efficient. Since each word unit inhibits every other word unit, the inhibitory input to a word unit *i* can be determined from the summed outputs of all the word units, less the output of word unit i. The inhibitory input to word unit i is then simply this difference times the word-level inhibition parameter *gamma[W]*.

On each sweep through the *wupdate* routine, the summed output of the word units from the previous cycle is used to determine the inhibitory input to each unit for the current cycle. At the same time, the output of all of the word units is accumulated for use on the next pass through the update routine.

RUNNING THE PROGRAM

The use of the **ia** program is much like the use of the other programs described in this book. Its main differences are the way inputs are indicated to the model and there are more parameters and more units than in most other models.

Trial and Forced-Choice Specifications

Trials are specified by entering the *trial* command to the **ia:** prompt; details for using this command are given in the "New Commands" section. A separate command, called *fcspec*, is used to specify the position and alternatives to be tested in the forced choice.

Screen Displays

The program has far too many units to display all their activation values on the screen at once. To compensate for this, the program keeps track of summary information about the activations in the network, as well as a *display list* of units for display to the screen.

The summary information consists of the current cycle number, the number of active units at the word level and in each letter position at the letter level, and the summed activation of all the active units at the word level and in each position at the letter level. An active unit is defined to be a unit whose activation is greater than 0.

The display list is a list of units whose activations are to be displayed. This list can be specified by the user, using the *get/ dlist* command. Alternatively, the program will compute its own display list at the end of each processing cycle. In this case, the display list consists of all of the units whose activation exceeds the values of the *dthresh/ word* and *dthresh/ letter* parameters, up to 15 letters in each position and up to 30 words.

New Commands

The **ia** program introduces only a small number of new commands:

fcspec
: Allows the user to specify a forced-choice test for the simulation, much as Reicher did in his experiments. The command first prompts for a position (enter 0, 1, 2, or 3) followed by a pair of letters, the first of which is the correct alternative and the second of which is the incorrect alternative. When a pair of choice alternatives has been specified, at the end of each cycle, the program will compute the probability that each would be chosen in a two-alternative forced-choice test.

print
: Allows the user to inspect the activations of each of the letter and word units and to inspect the response probability values associated with each letter unit. The command responds with a **print words?** prompt. If the user enters a string beginning with *y*, the program prints a screen full of words and their activations, and then presents the **p to push /b to break /<cr> to continue:** prompt. Responses of *p* and *b* have the usual effects; *return* causes the next screen of words and activations to appear. After finishing with the words, the program then prompts **print letters?** If the answer is yes, the activations of all of the letter units are presented for all four positions. Next, the command prompts **print letter resp-probs?** If the answer to this is yes, then for each letter, the probability that the letter would be given as the model's response in each position is displayed. Since this display would be overwritten by the top-level menu, the program prompts for a *return* before returning to the top level.

trial
: Allows the user to specify a sequence of *fields*, each containing an onset time and a field specifications. After the command is entered, the program presents a series of prompts of the form:

field #N: time:

Here *N* is the ordinal number of the field. To the first such prompt, the user enters the time for field 0, which is usually cycle 1, followed by the field specification. (A prompt for the field specification is provided if the cycle number is followed by a *return.*) The program then prompts for the next time-specification pair. When all the desired fields have been specified, enter *end* or type an extra *return*. Note that the times must be strictly increasing and that the time for field 0 must be 1 or greater. If the time for field *N* is less than or equal to the time for any preceding field, the program will never move on to field *N*. The last field encountered is just left on indefinitely, as is the case in most experiments. The field specification itself consists of a sequence of four characters. Allowable characters, and their meanings, are as follows:

Letters Letters are translated into the feature specifications of the corresponding uppercase letter as found in the Rumelhart-Siple font. Letters may be entered either in uppercase or lowercase.

Specifies the mask character, which is formed by turning on all the features that are present in either an *X* or an *O*.

? Question mark requests a random feature array; the random number generator is consulted to determine whether, for each feature, the presence or the absence unit should be on.

_ The underscore character is used for blanks; it requests that neither presence nor absence units be turned on in the corresponding position.

. The period character turns on all of the absence features and turns off all of the presence features; not to be confused with "_" above.

* The asterisk sets up the input array with the set of features common to the letters *K* and *R*, leaving both the presence and absence units off for the features that distinguish these two characters in the Rumelhart-Siple font.

" The double quote character informs the program that you wish to specify exactly which presence and which absence units should be turned on manually. You are prompted to supply these specifications immediately after entering the specification string containing this character. You are first prompted to specify which presence units should be turned on, then prompted to specify which absence units should be turned on. Each specification is given as an uninterrupted string of 0s and 1s. The indexes of the features in the Rumelhart-Siple font are as follows:

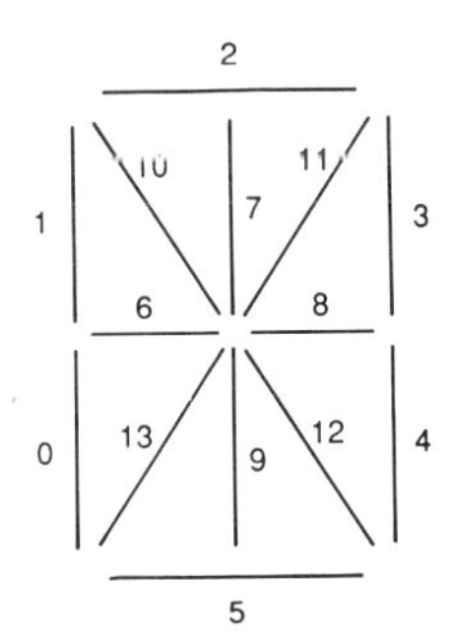

Thus, the letter *R* is designated by the presence specification:

11110010100010

and the absence specification:

00001101011101

disp/ opt/ lthresh
disp/ opt/ wthresh

Allow the user to set the minimum activation required for letters and words to be entered in the display list when the list is being generated automatically by the program; these variables have no effect when the display list is entered manually using the *get/ dlist* command. These commands can be useful when a large number of units are activated, particularly at the word level, because of the length limitations of the *dlists.*

get/ dlist

Allows the user to specify the display list or to clear one if a list has previously been specified. In response to this command, the program prompts:

enter words or – for dynamic specification:

Typing a "–" at this point will clear the old specification, if any, and return to the command level; if this is done, the program will construct a display list for each trial, as it does before a *dlist* is specified by the user. Entering *end* or typing *return* will set up a specification that specifies no words; any other strings are taken to be words. The program looks up the string in the list of word-unit names, and if the string is found, it is placed on the word display list. The program then prompts for subsequent words. The "–" option is no longer available at this point, but *end* or *return* can be given to end the list of words. After the end of the word list, the program prompts for letters for each of the four letter positions, starting with position 0. The list of letters for each position is terminated by *end* or an extra *return*. Due to limitations of screen space, the display lists are limited to 30 words and 15 letters per position.

Variables

The **ia** program has no configuration or environment variables; it has only one state variable, *cycleno*, and one mode variable, *comprp*. However, it has a large number of parameters. In fact, the parameters tend to be organized into parameter arrays, consisting of from two to as many as seven separately specifiable parameter values. For example, there are three different between-level excitation parameters. These are called *alpha/ f->l*, *alpha/ l->w*, and *alpha/ w->l*, for feature-to-letter, letter-to-word, and word-to-letter excitation, respectively. All of the *alpha* parameters are grouped together under the specifier *alpha*, under *set/ param/*. Thus, to set *alpha/ l->w* to 0.01, you would enter

set param alpha l->w 0.01

or, more compactly,

se p a l 0.01

The following list indicates all of the new or changed variables accessible through the *set* and *exam* commands. When there is a parameter array, the array name is followed by a slash and the names of the members are given in {} after the slash.

stepsize
: Determines the frequency of screen updating and/or pausing as in other programs. Available values are *cycle* and *ncycles*; by default the value is *cycle*.

mode/ comprp
: When set to 1, enables the computation of response probabilities for letter units. When set to 2, enables computation of response probabilities for word units as well. The default value is 1 since, typically, responses are assumed to be read from the letter level.

param/ alpha/ {f->l, l->w, w->l}
: Feature-to-letter, letter-to-word, and word-to-letter excitation parameters.

param/ beta/ {letter, word}
: Decay parameters for units at the letter and word levels.

param/ estr/ {p0-p3}
Relative strength of the external input in each letter position. These are set to 1.0 by default.

param/ fdprob/ {f0-f6}
For each field (f0-f6) specified in the trial command, the probability that each feature specified in the field specification is in fact detected. These are set to 1.0 by default.

param/ fgain
The scale factor for setting frequency-based resting activation levels for word units; larger values accentuate differences in resting levels as a function of word frequency. Defaults to 0.05.

param/ gamma/ {f->l, l->l, l->w, w->l, w->w}
Inhibition parameters for feature-to-letter, letter-to-letter, letter-to-word, word-to-letter, and word-to-word connections.

param/ max/ {letter, word}
Maximum activation parameters for letter and word units.

param/ min/ {letter, word}
Minimum activation parameters for letter and word units.

param/ orate
The rate of accumulation of the time-averaged activation used in computing response strengths.

param/ oscale/ {letter, word}
Multipliers used in scaling the time-averaged activations of units to obtain response strengths. Larger values produce larger differences in probability for a given difference in degree of activation.

param/ rest/ {letter, word}
Resting activation levels for letter-level and word-level units. For word-level units, this is the base resting activation; frequency-based offsets are multiplied by the *fgain* parameter and then subtracted from the base to obtain the true resting level.

param/ thresh/ {letter, w->l, w->w}
Output thresholds for letter-level and word-level units. The activation a unit passes to other units is equal to its activation minus its threshold, or 0, whichever is larger. For word units, there are separate values for outputs from words to other words and for output from words to letters.

OVERVIEW OF EXERCISES

The exercises we will propose here allow you to explore the behavior of the interactive activation model, using selected example stimuli. Generally, we have chosen the same items that were used in the examples described in McClelland and Rumelhart (1981) and Rumelhart and McClelland (1982).

Ex. 7.1. Using Context to Identify an Ambiguous Character

The first exercise gives you the opportunity to simulate the role of word-to-letter feedback in resolving ambiguity in visual input using the *R-K* example from McClelland and Rumelhart (1981). To run this exercise, get into the working directory you have set up for the **ia** program and enter

ia ia.tem ia.str

This sets up the display, and it results in the presentation of the **ia:** prompt along with a list of available menu options.

To set up the program to run this example, you must use the *trial* command to indicate the input to the program. To display the string *WOR** where the * stands for a character that is ambiguous between *R* and *K* enter:

trial 1 WOR end*

This indicates that at the beginning of cycle 1, the display designated by the string *WOR** should be presented. The word *end* indicates that no further fields are to be specified. Thus the display is turned on and left on indefinitely by this *trial* command. If you just enter *trial* by itself, the program will prompt you, first for a *field time* and then for *field contents*. Alternatively, you can enter everything on a single line as we have shown. (As in other programs, the *end* can be replaced with an extra *return*.)

Note that the display specification given here consists of uppercase letters and an asterisk. Lowercase letters may also be entered; they are treated as equivalent to uppercase. The "*" is one of several characters that have special meaning to the *trial* command, as described above; this one is specific to this particular example since it specifies the ambiguous *R-K* character, as shown in Figure 3.

FIGURE 3. The inputs specified by the field specification *WOR**. (From "An Interactive Activation Model of Context Effects in Letter Perception: Part 1. An Account of Basic Findings" by J. L. McClelland and D. E. Rumelhart, 1981, *Psychological Review*, *88*, 375-407. Copyright 1981 by the American Psychological Association. Reprinted by permission.)

Now that you have specified a display, you can begin the process of simulating the perception of the display. Simply enter *cycle*. The program will run for 10 cycles, since this is the number of cycles specified in the *ia.str* file.

The display that you will see at the end of 10 cycles is shown in Figure 4. At the top of the display area (below the command information) the current cycle number is shown, followed by the number of active units and their summed activation for each pool of letter units and for the word units. Below this information are three fields: two smaller ones, labeled TRIAL and CHOICE, and a larger, unlabeled one, which is used for displaying the activations of units on the *dlist*.

The *dlist* field displays the activations and response probabilities of units on the *dlist*. Since no fixed *dlist* has been specified, the units that are displayed are those whose activations are greater than the values of the display options *lthresh* (for letters) and *wthresh* (for words). There is a separate column for each letter position, and room for two columns for words, though in this instance only one of these is used. Within each of these columns there are three subcolumns. The first contains an identifier for each item, the second indicates its activation (in hundredths, so that 0.15 is given as 15 and 0.07 is given as 7), and the third gives the response probabilities.

An exception is made in the case of the word response probabilities. These take a tremendous amount of time to compute, and since for the moment we are focusing on context effects on letter perception and we are not concerned with word response probabilities, these are neither computed nor displayed. (If you wish to compute and display these probabilities, set the *comprp* mode variable to 2 and the *dlevel* to 3.)

```
ia:
disp/  exam/  get/  save/  set/  clear  cycle  do  fcspec  log  newstart  print
quit  reset  run  trial

cycle  10    letter 0    letter 1    letter 2    letter 3     word
             num     1   num     1   num     1   num     2   num       5
             sum    62   sum    62   sum    62   sum    71   sum      52
------------------------------------------------------------------------
  TRIAL    | w  62 26    o  62 25    r  62 26    k  44 15    word  8
  1 WOR*   |                                     r  27 12    wore  6
           |                                                 work 32
           |                                                 worm  0
           |                                                 worn  4
           |
           |
           |
-----------|
 CHOICE    |
pos      0 |
-    0   0 |
-    0   0 |
fca      0 |
```

FIGURE 4. The state of the **ia** program after 10 cycles processing *WOR**.

Since both *lthresh* and *wthresh* default to 0, the items whose activations exceed 0 are displayed in alphabetical order until the available display space runs out; if the activation is between 0.00 and 0.01, a 0 is displayed. The display area allows 15 letters in each position and 30 words. Thus, if more than 30 words are activated (as indicated in the *word num* field in the upper portion of the display), only the activations of the 30 words that come earliest in the alphabet will be displayed in the *dlist* display. If this ever happens in a simulation, you might want to increase the value of *wthresh*, which is accessed by the *disp/ opt/ wthresh* command.

It is worth reiterating the meaning of the response probabilities. These numbers are computed using the formulas given earlier in the "Readout" section under the description of the model; they can increase or decrease through the course of processing. What these numbers represent at any given moment is the probability that the corresponding item *would* be chosen as a response, *if* readout were to occur at the present moment. Readout occurs at the end of processing when the display is degraded and is not followed by a mask; for trials in which the target display is followed by a mask, we assume readout occurs at the optimal time. This is fairly constant across different displays and is mostly affected by the timing of the onset of the mask field.

To the left of the main display area is the trial field. This simply indicates the onset times and contents of the fields that have been specified in the *trial* command. Below this is the choice field, which is not relevant in this exercise.

To summarize what can be seen on the screen after 10 cycles, only one letter is activated in each of positions 0, 1, and 2. In position 3 (the last position) the letters *k* and *r* are both active; *k* is beginning to get stronger than *r* because of feedback from the word level. At the word level, we can see that the words *word, wore, work, worm*, and *worn* are all active to some degree, although *work* is clearly most active.[1]

Now that the display has been explained, reinitialize the program (*reset* and *newstart* are equivalent for this exercise because there is no element of randomness), set *single* to 1, and step the program along, observing on each cycle the build-up of activation at the letter and word levels. Run about 40 cycles.

Q.7.1.1 When does *k* start to gain an edge over *r* in activation? Why? Even though *k* becomes more active, *r* is not suppressed. What parameter controls this? Explain why the response probability for *r* eventually does go down, even though its activation does not. Also explain why the response probabilities for *r* and *k* tend to pull apart more slowly than their activations.

---

[1] We will follow the convention of indicating visual displays and choice alternatives in uppercase and indicating unit names and activations in lowercase. Thus the display *WOR** activates the word unit *work*.

Ex. 7.2. Simulations of Basic Forced-Choice Results

In this exercise we will give you the opportunity to simulate some basic experimental phenomena that we considered in McClelland and Rumelhart (1981); in particular, we'll examine the forced-choice advantage of letters in words over letters in unrelated context and the fact that the word advantage holds for letters in pronounceable pseudowords as well as actual words.

First simulate the advantage of letters in words over letters in unrelated context, using as your example the letter *V* in *CAVE* compared to the letter *V* in *ZXVJ*. Assume that the target display is presented at the beginning of cycle 1, followed by a mask containing a sequence of mask characters, appearing at the beginning of cycle 16. Assume that the forced choice tests for the letter *V* in position 2 (counting from 0) and that the incorrect alternative is the letter *G*.

The trial specification for *CAVE* is

trial 1 CAVE 16 #### end

The string #### specifies a string of mask characters (each consisting of an *X* superimposed on an *O*). For the *ZXVJ* trial specification, simply replace *CAVE* with *ZXVJ*.

For this exercise, we introduce the *fcspec* command and the choice field of the display, which has thus far remained blank. To set up the choice field, you must specify a letter position to test and a pair of choice alternatives using the *fcspec* command. For this exercise, the position to test is position 2 (since we count from 0), the correct alternative is *V*, and the incorrect alternative is *G*, so the following command specification should be entered:

fcspec 2 V G

You will now see the choice field in the lower left-hand corner of the screen filled in with the information you have specified. This field contains:

- The letter position specified in *fcspec*, if any (defaults to 0).
- The correct alternative (or "–" if none has been specified), its activation, and its current choice probability.
- The incorrect alternative (or a "–" if none), its activation, and its current choice probability.
- The forced-choice accuracy (labeled *fca*) that would result if processing were terminated at this point.

The *current choice probability* is the probability that the alternative would be chosen if the forced choice were to be based on readout of letter activations on the current cycle. The *forced-choice accuracy* is the maximum value thus far attained for the current choice probability of the correct alternative. The value of this number at the end of the trial is taken to be the probability that the model would make the correct forced-choice response.

To run these examples, you will probably want to run about 20 cycles for each item. By cycle 16, the response probabilities will have peaked, so additional cycles will have no effect on the forced-choice accuracy, but it is sometimes interesting to watch letter- and word-level activations begin to drop off again as the mask takes effect.

For this simulation it is also useful to specify a *dlist* so that you can see the activation and response probability for the incorrect choice alternative as well as for the correct response. The file *cave.dli* is useful in this regard. It contains a *get/ dlist* command that specifies that the *dlist* should contain *cave* plus all the other words that the model knows that have three letters in common with *CAVE*, and that the letters displayed should be *C* in position 0, *A* in position 1, *V* and *G* in position 2, and *E* in position 3. You can leave the same *dlist* in place for the *ZXVJ* example since this one activates no words. This command file is read in by using the *do* command:

do cave.dli 1

(the 1 indicates that the commands in this file are to be executed just once).

Finally, the parameters you need to be aware of to understand the word-level activations are the *alpha* and *gamma* parameters given previously in Table 1.

Q.7.2.1. Describe the time course of activation of the unit for *V* in position 2 in each of the two cases, and explain how this is translated into the forced-choice advantage for *V* in *CAVE* over *V* in *ZXVJ*. Also, consider what happens at the word level. Why does the string *ZXVJ* produce no word-level activations, given the parameters of the model? Why do no word units other than *cave* achieve substantial activation values when *CAVE* is presented?

Next we consider the processing of *pseudowords*—pronounceable nonwords such as *MAVE* or *REAT*. Experimentally, subjects perform almost as well with these stimuli as with real words, and they show a considerable advantage over single letters and letters in unrelated strings (Baron & Thurston, 1973; Johnston & McClelland, 1977).

At first glance, we might not expect perceptual facilitation with pseudowords in the model, since it contains units only for real words. However, as we shall see, the model does in fact replicate the pattern of human performance. To see this, compare the model's performance on the *V* in

MAVE to the results you have already obtained with *CAVE* and *ZXVJ*. Keep everything else the same as before.

The file *mave.dli* contains a useful *dlist* specification for this simulation. It sets up the letters *m,a,v,e*, the alternative *g*, and all the words that have three letters in common with *MAVE*. Read this file in with the command *do mave.dli 1* followed by *reset*. During the simulation runs, study the pattern of activation at the letter and word levels.

Q.7.2.2. Explain why forced-choice accuracy for *V* in *MAVE* is almost as good as for *V* in *CAVE*.

Q.7.2.3. Study the word-level activations that are produced when the item *MAVE* is shown. See if you can explain why the initial advantage of *have* is amplified through the course of processing so that by cycle 15 it is considerably more strongly activated than *gave* and *save*. Also, try to explain why *move* shows less activation than *save*, even though they have about the same frequency and therefore start out at the same resting level.

Hints. The fact that *have* is more active than *gave* or *save* after 15 cycles is influenced by the resting level; however, your job is not to explain simply why *have* starts out a bit higher; you have to say why its advantage is amplified.

Q.7.2.4. Choosing another pronounceable nonword, study the pattern of activation it produces and see if you can observe word-level effects similar to those you observed with *MAVE*.

Ex. 7.3. Subtler Aspects of the Word Superiority Effect

This exercise allows you to go beyond the basic word and pseudoword superiority effects, going more deeply into the accounts offered by the interactive activation model for some of the detailed findings reported in the literature on letter and word perception. In both parts of the exercise you will see that it often requires a careful consideration of the inner workings of the model as well as the characteristics of particular experiments to understand why particular results were obtained. The model serves not only to explain what often looks like a confusing pattern of results, but also helps pinpoint which aspects of the experiments are responsible for the findings obtained.

Bigram frequency effects. One surprising finding in the word superiority effect literature is the apparent absence of effects of bigram frequency on the size of the word superiority effect obtained in the Reicher paradigm

(McClelland & Johnston, 1977). To test for an effect of bigram frequency, McClelland and Johnston examined forced-choice accuracy of letters in high bigram-frequency words (such as *PEEL-PEEP*) and pseudowords (such as *TEEL-TEEP*) and in low bigram-frequency words (*POET-POEM*) and pseudowords (*HOET-HOEM*). They found a slight advantage for words over pseudowords, but no effect of bigram frequency either for the word or for the pseudoword stimuli. These results are particularly surprising in view of the fact that other studies have found effects that can be interpreted as supporting the notion that items of higher bigram frequency lead to more accurate perception (Rumelhart & McClelland, 1982, Experiment 9). Here we explore what light the interactive activation model can shed on these findings.

To see what happens in the interactive activation model with stimuli of the type used by McClelland and Johnston, run the **ia** program with the items *TEEL* and *HOET*, testing the last position (position 3) in each case, with *P* as the incorrect alternative on *TEEL* and *M* as the incorrect alternative on *HOET*. Set up the trial specifications as in the previous exercise, so that the letter string is displayed on cycle 1 with the mask coming on cycle 16.

Q.7.3.1. Describe the results you obtain in these tests, and explain why the forced choice comes out actually slightly favoring the low bigram-frequency item.

As the examples used above illustrate, McClelland and Johnston made up their stimuli according to the following procedure. They first constructed a set of high bigram-frequency word pairs like *PEEL-PEEP* and a set of low bigram-frequency pairs like *POEM-POET*. Then they derived high and low bigram-frequency pseudoword pairs from the word pairs by changing one of the context letters from each pair. Thus *TEEL-TEEP* was derived from *PEEL-PEEP* and *HOET-HOEM* was derived from *POET-POEM*.

Q.7.3.2. Try to explain why McClelland and Johnston's procedure for constructing pseudoword pairs by changing nontarget letters in word pairs may have prevented them from finding a difference between high and low bigram-frequency items.

Hints. You might consider what would happen if the pseudowords *HOEM* and *HOET* were used, testing performance on the *H* rather than the final letter. You can use *L* as the forced choice alternative.

In their Experiment 9, Rumelhart and McClelland (1982) report results that look at first sight to be inconsistent with those of McClelland and Johnston (1977); that is, Rumelhart and McClelland found that forced-choice performance on pseudowords did depend on a measure very closely related to bigram frequency. However, the materials used by Rumelhart and McClelland were not constructed from word pairs at all. Rather, all possible pseudoword items were constructed and separate sets of pairs were made from those having high and low letter-transition probabilities (a measure highly correlated with bigram frequency). For these items, both the experiment and the simulation obtained an advantage for the high transitional probability items.

Effects of contextual constraint. In the 1960s and 1970s, several studies were carried out examining the role of contextual constraint in word recognition. Contextual constraint refers to the degree to which the context a target letter occurs in restricts the possible identity of the target letter, based on which letters form words with the context. Thus _*LUE* is a relatively constraining context, because only three letters (*B*, *C*, and *G*) make words in this case;[2] whereas _*AKE* is much less constraining, since 10 letters can fit with it to make a word. The results of the experiments (most importantly, those of Broadbent and Gregory, 1968, and Johnston, 1978) presented a fairly complex picture. Effects of contextual constraints appear to be obtained under conditions in which subjects are shown a visually dim or degraded display with no patterned mask and are asked to give a free report. However, these effects do not appear when subjects receive a brief, masked presentation of the target word followed by a forced choice. The interactive activation model appears to explain this pattern of results.

This section describes how you can explore the account of these effects of contextual constraint using the **ia** program. Before we proceed further, however, we will note that these simulations are rather computationally intensive. Users who do not have a floating-point coprocessor and/or a lot of time may want to skip this section.

For these simulations, we will use the target words *CLUE* and *CAKE*. The forced choice (where applicable) will be applied to the letter *C*, with *B* serving as the incorrect alternative. The example is carefully chosen: *CLUE* and *CAKE* have about the same frequency and therefore the same resting level in the model. Forced-choice performance is modeled as before. For the free report, it is assumed that subjects read out their response from the word level, timing the readout to achieve optimal performance.

To begin examining the effects of contextual constraint, present the items *CLUE* and *CAKE*, using the same display conditions we have been using, in which the word is presented at the beginning of cycle 1 and is followed by the mask at cycle 16. Compare the forced-choice accuracy for both items, as well as the response probability for each target word at the word level. (To examine word-level response probabilities, you will need

---

2 We leave out the low-frequency word *FLUE*, which is not in the word list used by the IA program.

to set the *comprp* mode variable to 2 and the *dlevel* variable to 3.) Note that response probabilities build up very slowly at the word level and depend on the persistence of activations of words beyond the onset of the mask, so you will want to run about 32 cycles of processing to get a good sense of the probability of identifying the target word based on readout from the word level.

Now repeat the above experiment, using display conditions intended to simulate a visually dim or degraded display with no patterned mask. To do this, use the command

set param fdprob f0 0.65

to set the probability of feature detection in field 0 of the trial sequence to 0.65. With this setting, the model will only detect 65% of the features in the display, simulating the effects of visual degradation. Now give the trial specification, simply presenting the target string at cycle 1. For example, for *CLUE* the specification would be

trial 1 CLUE end

Run several runs with each target, letting processing go on for 50 cycles so that word-level activations and response probabilities reach asymptotic levels. Use the *newstart* command to reinitialize between runs to obtain a new random set of features on each run.[3]

Q.7.3.3. Try to explain why effects of contextual constraint emerge in the free report with dim target conditions, and why they do not occur with the forced choice under bright-target/pattern-mask conditions.

Ex. 7.4. Further Experiments With the Interactive Activation Model

This set of exercises is based on material from Rumelhart and McClelland (1982). The first part explores a prediction of the model that was verified by an experiment reported in that paper. The second part applies the model to a variant of the word superiority effect called the *contextual enhancement effect*. The last part considers how expectations might

---

[3] Actually, after a few runs you will begin to get a feel for what happens with the response probabilities. To speed things up, you can set the *comprp* mode variable to 1 or even 0 (eliminating computation of letter-level response probabilities) and just watch the activations of the units.

influence processing. These exercises are rather specialized and are provided for readers with a particular interest in further exploration of the IA model and the phenomena of word perception.

Facilitation effects with unpronounceable nonwords. In Rumelhart and McClelland (1982), we predicted that the perceptual facilitation for letters in pronounceable nonwords should also occur with letters in a class of unpronounceable nonwords that were made up specifically to have several "friends"; that is, that were made up so that they had three letters in common with several words, even though they were not pronounceable. An example is *SLNT*, with a forced choice between *L* and *P*. As predicted, we found roughly the same forced-choice advantage for such items as for items like *SLET*, compared to random strings like *JLQX*.

In this exercise, explore the model's handling of the items *SLNT* and *SLET* compared to *JLQX*, using the display conditions of Ex. 7.2. Then consider the following question:

Q.7.4.1. Why does the model predict roughly equal facilitation for the *L* in *SLNT* as for the *L* in *SLET*? What characteristics of the items appear to be critical? Do simulations with other selected items to see if you are correct.

The contextual enhancement effect. Rumelhart and McClelland (1982) reported a phenomenon called the contextual enhancement effect. This is the finding that letters are perceived better when the *context* they are in is enhanced by presenting it for a longer time. This effect is demonstrated in experiments using trial sequences in which the context is presented before the target letter is shown, as in this trial specification:

```
 1 _HIP
13 SHIP
25 ####
```

Here the subject is given a forced choice between *S* and *W* for letter position 0. Forced-choice accuracy is greater when the context is presented before the full target field than in a control condition in which there is no preview of the context:

```
 1 SHIP
13 ####
```

The effect can be obtained both with words and with pronounceable nonwords like *SHIG*. In the experiments, the context boosted accuracy about 8% for both kinds of materials; there was an advantage for words, both with and without a preview of context, of about 4%.

For this exercise, simulate the context enhancement effect, with both *SHIP* and *SHIG*. That is, run trials with and without a preview for both *SHIP* and *SHIG*.

Q.7.4.2. Describe how well the model does at accounting for the contextual enhancement with words and pseudowords. See if you can explain the results that you observe.

In answering Q.7.4.2 you will have discovered that the model does not produce an enhancement effect with the pseudoword *SHIG* when the default parameters are in effect. This item is typical of most pseudowords. After considerable exploration, we were able to find a new set of parameters that did allow us to accommodate the context enhancement effect with pseudowords. The file *pwcee.par* contains commands that will produce the required modifications to the parameters. These commands can be executed using the command *do pwcee.par 1*. You may use these parameters, or (if you really want a challenge) you may come up with a set of your own that works. (As a hint, we will note that the modifications that are necessary have more to do with resting levels and thresholds than with excitation and inhibition parameters.)

Q.7.4.3. Using either the parameters in *pwcee.par* or your own parameters, simulate the context enhancement effect with pseudowords. Explain why the modifications work.

Expectation effects. Carr, Davidson, and Hawkins (1978) reported an interesting finding on the role of subjects' expectations in perceiving letters in words, pseudowords, and random-letter strings. They set up their subjects to expect either words, pseudowords, or random letters to be presented, and then, using a small number of trials of other types surreptitiously included in their materials, they were able to assess the advantage for words and pseudowords in each of these three expectation conditions. Their results are shown in Table 2.

Basically, what they found is that there was a word advantage over random letters, regardless of the subject's expectations, but that there was a pseudoword advantage over random letters only if the subjects were actually expecting pseudowords to be shown. Try to figure out for yourself whether the interactive activation model can account for these results, assuming that subjects are able to exert strategic control over one or more parameters of the model.

In simulations for this exercise, it is sufficient to use a single word, a single pseudoword, and a single random letter item to study these effects; for example, *CAVE*, *MAVE*, and *ZXVJ* will do. Use the standard parameters and set up trial specifications with the mask coming at cycle 16 as usual.

TABLE 2

EFFECT OF EXPECTED STIMULUS TYPE ON THE WORD AND PSEUDOWORD ADVANTAGE OVER RANDOM LETTERS

	Expectation		
Target	Word	Pseudoword	Random letters
Word	0.15	0.15	0.16
Pseudoword	0.03	0.11	−0.02

Note. From "Perceptual Flexibility in Word Recognition: Strategies Affect Orthographic Computation but not Lexical Access" by T. H. Carr, B. J. Davidson, and H. L. Hawkins, 1978, *Journal of Experimental Psychology: Human Perception and Performance, 4,* 674-690. Copyright 1978 by the American Psychological Association. Reprinted by permission.

Q.7.4.4 Experiment with different parameters of the model and see which ones can produce the effects of Carr et al. Describe each change you made, indicate how well it worked, and tell why it had the effects that it had.

Our answer to this question involved control over one of the excitation or inhibition parameters (we will not say which). Perhaps if you change other parameters you could obtain some or all of the same effects.

APPENDIX **A**

Setting Up the PDP Software on the PC

In this appendix we describe the procedure for unpacking the PDP software for use on PCs and other MS-DOS computers. We first describe how the files are organized on the two diskettes. Then we explain how to set up working directories and how to unpack the software from the diskettes and make it ready for use. The discussion assumes some experience with MS-DOS. To unpack a single program, you can skip directly to the end of this appendix, where a script is given.

Organization of the Diskettes

Each diskette contains a number of *archives*. There is one archive for each simulation program, one archive called *utils.arc* for the **plot** and **colex** utility programs, and one archive called *src.arc* for the source files. The name of each archive file ends with the *.arc* extension. Disk 1 contains *iac.arc*, *cs.arc*, *pa.arc*, *bp.arc*, and *aa.arc*. Disk 2 contains *cl.arc* and *ia.arc*, as well as *utils.arc* and *src.arc*.

Each archive consists of a number of files that have been compressed and then stored together by a program called ARC. Each program archive contains the PC-executable version of the corresponding program, as well as all the pattern, network, start-up, template, look, and weight files that may be needed to run the examples described in the various chapters of this book. The *utils.arc* archive consists of the executable versions of **plot** and **colex**. The *src.arc* archive consists of source files for each executable program and for the routines that make up the various general routines described in Appendix F, as well as files containing commands that allow these programs to be recompiled, as described in Appendix G. Appendix G also describes recommended procedures for users who wish to port the software to UNIX systems.

Setting Up Working Directories

Generally, it makes sense to set up a working directory for each simulation program. If you have a system with no hard disk, you will have to create these on a floppy. As a rough guideline, a double-sided, double density 5¼" floppy disk will hold two programs comfortably. Thus, you might put directories for the **iac** and **cs** programs on one floppy, **pa** and **bp** on another, **aa** and **cl** on another, and **ia** on a floppy by itself. The **plot** and **colex** programs, if you find you wish to use them, should normally be placed either in the directory where you wish to use them or in a directory that MS-DOS will check in looking for executable programs. The source files take up rather a lot of space and may have to be placed on a disk by themselves.

It is a good idea to have a working copy and a back-up copy of all of the software so that if anything happens to either copy you have the other. From this point of view it makes sense to set up working directories for all the programs right away, and then to put away the distribution disks for safekeeping. Alternatively, you might simply want to use the MS-DOS DISKCOPY utility to create back-up copies of the distribution floppies, and unpack archives as you need them.

If you have a hard disk, it makes sense to place each program in a subdirectory of the the same parent directory. For example, you might create a parent directory called *pdp*, with subdirectories called *iac*, *cs*, and so on.

Working directories are created using the MS-DOS *mkdir* command. For example, to make a directory called *iac* on a floppy disk in drive B you would enter

mkdir b:\iac

or you could connect to drive B by entering

b:

and then simply enter

mkdir iac

Of course, the floppy must be formatted before the *mkdir* command will work. See your MS-DOS documentation for formatting instructions if necessary.

Once you have created the appropriate directories, unpacking the software is extremely simple. You place the diskette containing the archive you wish to unpack in drive A of your PC. Make sure the target directory is in place either on your hard disk or on a floppy installed in drive B. Enter

a:

so that your working directory becomes the top level directory on drive A. Then enter

arce arcname destdir

where *arcname* is the name of the archive you wish to extract files from and *destdir* is the name of the directory you wish to place the files in.

For example, to extract all the files from *iac.arc* and place them in a working directory called *iac* on a floppy disk in drive B, you would enter

arce iac b:\iac

To extract all the files from *src.arc* and place them in a directory called *pdp\src* on your hard disk (drive C), you would make sure disk 2 was in drive A, and then you would enter

arce src c:\pdp\src

Once you have set up a working directory for a program, you may move to that directory with the *cd* command and begin to work. Thus, to use the **iac** program, assuming you have unpacked it to a directory called *iac* on a floppy in drive B, you would enter

b:
cd iac

Note that **arce** is a simple version of the ARC program that is only able to extract files from archives. Some further information about it can be obtained by entering *arce* when connected to the drive containing either of the two floppies.

A Script to Unpack a Single Program

The following sequence of actions will set you up to run the **iac** program. If you wish to set up a different program, just use its name instead of **iac** in what follows.

1. Put a formatted floppy disk into drive B.

2. Create a directory for the **iac** program by entering

   mkdir b:\iac

3. Put disk 1 of the PDP software into drive A.

4. Connect to drive A by entering

   a:

5. Unpack the files by entering

   arce iac b:\iac

6. Connect to drive B by entering

   b:

7. Change to the *iac* directory by entering

   cd iac

APPENDIX **B**

Command and Variable Summary

In this appendix we give a brief description of all of the commands available in the PDP programs. With each command we indicate the program or programs the command is relevant to and the number of the page where the command was described. Where there are different variants, we describe these and indicate where each variant was described. Since this summary includes all of the subcommands of the *set/* command, it also provides a summary of all of the variables that are accessible in the programs.

clear — all [30]

Clears the display area.

ctest — **aa** [171]

Completion test. Used to see how well an auto-associator can fill in cleared parts of a specified pattern on the pattern list. Prompts for pattern name or number and elements to clear for completion testing.

cycle — **iac, cs, bp, ia** [30]

Runs *ncycles* of processing. In **bp**, applies only in *cascade* mode.

do — all [30]

Executes the commands in a file a specified number of times. Prompts for the file name and an integer number of repetitions.

fcspec **ia** [223]

Forced-choice specification for testing letter perception in the interactive activation model. Prompts for a letter position (0-3), a correct alternative, and an incorrect alternative.

input **cs, iac** [30]

Allows user to specify external inputs to units by name or number. First asks whether old specifications should be cleared (indicate *y* or *n*) then prompts for unit names or numbers, followed by an external input value. The value specified is scaled by the parameter *estr*. To terminate input, enter *end* or press *return*.

log all [30]

Used to open or close a log file. Prompts for a file name or "–". If a file name is entered, the existing log file is closed if there is one, and a new one is opened with the specified name. If "–" is entered, any existing log file is closed.

newstart **cs, ia** [56]

Allows the program to be retested on the same problem, but with a new series of pseudorandom numbers. Chooses a new random seed for the random number generator then reinitializes the activations of units in the network.

pa, bp, aa, cl [104]

For these programs, *newstart* also reinitializes the weights in the network. In **pa** and **aa** they are reset to 0s, in **cl** they are reset to random initial values, and in **bp** they are reinitialized as specified in the *.net* file.

print **ia** [223]

Prints activations of all word and letter units and response strengths of all letter units on the screen in the interactive activation model.

ptrain **pa, bp, aa, cl** [104]

Permuted training. Executes *nepochs* of training, presenting each pattern or pattern pair that is on the pattern list once in each epoch. A new random order of presentation is used on each epoch.

quit all [31]

Quits the program. Prompts for confirmation. Enter *y* to quit, anything else to cancel the command.

reset **iac** [31]

Resets activations of units in the network to resting level.

cs, ia [56]

Allows the program to be retested on the same problem, with the exact same pseudorandom number sequence used to determine order of *rupdate* in **cs** and other random aspects of processing. Reinitializes the random number generator to the value of the *seed* variable, which will be the same as the value used when the process was last initialized unless *seed* has been set manually to a new value.

pa, bp, aa, cl [104]

For these programs, *reset* also reinitializes the weights in the network. This means that for **bp** and **cl**, the same random starting weights that were used on the previous run will be used again. *Note:* The starting weights used for Ex. 5.1 must be read in from the file *xor.wts* after the reset command is entered, using the *get/ weights* command.

run all [31]

Used to pass a command to the command interpreter under which you are running the simulation program. The command to be run is terminated by *end* or an extra *return*. Thus, to see if a file exists in your directory named *foo.wts*, you could enter *run dir foo.wts end.*

strain **pa, bp, aa, cl** [104]

Sequential training. Executes *nepochs* of training, presenting each pattern or pattern pair on the pattern list once in each epoch. Patterns are presented in the same fixed order on each epoch, as specified in the *.pat* file.

tall **pa, bp, aa, cl** [104]

Tests all patterns on the pattern list. Patterns are presented for testing one at a time, and the program pauses after each pattern to allow you to examine the results. Learning is turned off while *tall* is in progress.

test **iac, cs, bp, cl** [31]

Presents a pattern from the pattern list for testing. Prompts for the name or number of the pattern.

pa [105]

Allows testing of a pattern from the pattern list, a distortion of such a pattern, or a specific pattern as entered from the keyboard, as indicated by user's response to the prompt. The pattern is tested against a target pattern which may either be one of the targets from the pattern list, a distortion of such a target, or a specific pattern entered from the keyboard. If the user chooses to enter a pattern from the keyboard, the pattern must be terminated by *end* or an extra *return*.

aa [171]

Allows testing of a pattern from the pattern list, a distortion of such a pattern, or a specific pattern entered from the keyboard; also allows retesting of the last pattern tested. This is useful for studying the effects of a single training experience on a specific pattern. A pattern entered from the keyboard must be terminated by *end* or an extra *return*.

trial **ia** [223]

Allows user to set up a processing trial consisting of a sequence of displays in the interactive activation model. Each display consists of an onset time followed by a string of four characters indicating the contents of the display. If the characters are letters, they specify the corresponding feature patterns in the Rumelhart-Siple font. Other special characters are used for other feature patterns, as described in Chapter 7.

disp/ all [31]

Header for various specific commands that allow users to display information on the screen or to adjust options associated with displaying information to the screen.

disp/ state all [31]

Display the current state of the network. Clears the screen and then redisplays the current states of all of the templates that should be visible, given the current value of the *dlevel* variable.

disp/ <template> all [31]

Display *<template>*, where *<template>* is the name of one of the templates defined in the *.tem* file.

disp/ opt/ all [31]

Header for various commands that allow use to alter different display characteristics.

disp/ opt/ lthresh **ia** [225]

Allows user to set the minimum activation required for letters to be entered on the display list in the **ia** program. Only letters in the display list are displayed when the screen is updated. The *lthresh* is ignored when an explicit display list is provided by the user.

disp/ opt/ standout all [31]

Allows the user to set the value of the *standout* option for displaying negative numbers. If *standout* is set to 0, minus signs are used to indicate negative numbers where possible. When *standout* is set to 1, as it is by default, reverse video is used for negative numbers.

disp/ opt/ wthresh **ia** [225]

Allows user to set the minimum activation required for words to be entered on the display list in the **ia** program. Only words in the display list are displayed when the screen is updated. The *wthresh* is ignored when an explicit display list is provided by the user.

disp/ opt/ <template> all [32]

Allows the user to set various options associated with the specified *<template>*, where *<template>* is the name of one of the templates defined in the *.tem* file.

exam/ all [32]

A synonym for *set/*; it allows the user to examine the current value of a variable and to set it to a new value if desired. The subcommands available under both *set/* and *exam/* are described below under *set/*.

get/ all [32]

Header for commands that allow the user to specify lists of values with a single command or to specify a file from which to read a complex specification such as the network specification.

get/ annealing **cs** [56]

Used to specify an annealing schedule for *boltzmann* and *harmony* modes in the **cs** program. The user is prompted for an initial temperature, then for a sequence of time-temperature pairs. The program interpolates linearly between these milestones, and stops adjusting temperature at the last milestone. Follow the last time-temperature pair with *end* or type an extra *return*.

get/ dlist **ia** [225]

Used to specify a display list specifying which letter and word units will be displayed when the screen is updated in the **ia** program. The user is prompted for words and letters to place on display list.

get/ network **iac, cs, pa, bp** [32]

Used to tell the program to read a network specification file and set up the network of units according to its contents. Prompts for the name of the network specification file.

get/ patterns **iac, cs, aa, cl** [32, 171]

Used to tell the program to read a list of patterns from a pattern specification file. Each pattern specification is a sequence of entries consisting of a pattern name (a string of characters beginning with a nondigit) followed by a sequence of *ninputs* floating-point numbers corresponding to the elements of the pattern. Entries in the pattern file are separated by white space (spaces, tabs, or newlines).

pa [105]

As above, except that in **pa**, the program expects the pattern specification file to contain pattern pairs. Each pair has a sequence of entries consisting of a pattern name, followed by *ninputs* floating-point numbers corresponding to the elements of the input pattern, followed by *noutputs* numbers corresponding to the elements of the target pattern.

bp [142]

As with **pa**, with special handling of negative input and target entries. A negative input entry specifies that the input element should be the previous activation of the unit indexed by the absolute value of the entry. A negative target entry specifies a "don't care" element; that is, that no target value is to be specified for the corresponding output unit.

get/ rpatterns **aa** [172]

Used to generate a list of random patterns of +1s and −1s for testing in the auto-associator. Prompts for a number specifying the number of patterns to construct, and then prompts for a number specifying the probability that each element will be +1. Patterns are named *rN*, where *N* is an integer from 0 to *npatterns* − 1.

get/ unames all but **ia** [33]

Allows the user to specify a list of names for the units in the network. Each name is a sequence of characters, and successive names are separated by spaces, tabs or newlines. Follow the last entry with *end* or type an extra *return*.

get/ weights all but **ia** [33]

Used to tell the program to read a file containing weights for all of the connections in the network. For networks that have bias terms or sigma terms (this applies to *harmony* mode in **cs** only), these are also read in from the same file. Prompts for the name of the weight file. See Appendix C for further information about the format of the weight file.

save/ all [33]

Header for commands that allow user to tell the program to save various kinds of things in a file.

save/ patterns **aa** [172]

Allows the user to save the patterns on the pattern list in a file, together with their names, so that they can be read in later using the *get/ patterns* command. Prompts for the name of the file.

save/ screen all [33]

Allows the user to save the current screen display in a file. Redisplays the screen exactly as it would look in the file, then prompts for the name of the file.

save/ weights all but **ia** [33]

Allows the user to save the current values of the weights (and bias or sigmas if any) in a file, so that they can be read in again later using the *get/ weights* command. The format of the file is described in Appendix C. The command prompts for the file name.

set/ all [33]

Allows the user to examine the current value of a variable and to set it to a new value if desired. The commands *set/* and *exam/* are synonyms. The following list indicates the subcommands available under *set/*; all of these are in fact the names of variables. Variables whose names are followed by "*" are installed when the network is initialized, and so will not be visible before the program is initialized (e.g., right after the program is started up and before the *get/ network* command has been executed). For vector variables, an index must be given after the variable name to specify which entry is to be accessed. For matrix variables, two indexes must be given. The first corresponds to the row of the matrix and the second to the column.

*set/ bias** **bp, cs, pa** [57]

Vector of bias parameters associated with the units in the network.

set/ dlevel all [35]

The global display level variable. When the display is updated, all templates with display levels less than or equal to the global display level are displayed.

set/ ecrit **aa, bp, pa** [106]

Criterion value for the total sum of squares *tss* used to determine when to stop training. Training stops when the *tss* falls below the value of *ecrit*.

set/ lflag **pa, bp, aa, cl** [106]

A flag variable indicating whether learning (connection strength adjustment) should take place or not. Learning occurs when *lflag* is not equal to 0.

set/ ncycles **aa, bp, cs, ia, iac** [35, 142]

The number of processing cycles executed when the *cycle* command is entered or when the *test* command is issued. In **bp**, applies only in *cascade* mode.

set/ nepochs **pa, bp, aa, cl** [106]

The number of epochs of training that are carried out when the *ptrain* or *strain* commands are issued.

set/ nupdates **cs** [57]

The number of processing unit random updates that are executed in each processing cycle in the **cs** program. Except in *harmony* mode, *nupdates* is generally set equal to *nunits* so that each unit is updated an average of one time per cycle.

set/ seed all [36, 57]

The current value of the seed that is used when the random number generator is reinitialized. Note that the *newstart* command assigns a new random value to *seed* before reinitializing the random number generator.

*set/ sigma** **cs** [57]

Vector of sigma values associated with units in *harmony* mode of the **cs** program.

set/ single all [36]

A flag variable that determines whether the program will pause or not after each processing step, where the size of the step is determined by the *stepsize* variable defined below. Pausing occurs if *single* is nonzero.

set/ slevel all [36]

The global save level variable. When the display is updated, all templates with display levels less than or equal to the global *dlevel* variable and less than the global *slevel* variable are logged in the log file, if one is open.

set/ stepsize **iac** [36], **cs** [57], **pa** [106]
bp [142], **aa** [172], **cl** [197], **ia** [226]

The size of the processing step between display updates and pauses. The recognized values are *update*, *cycle*, *ncycles*, *pattern*, *epoch*, and *nepochs*.

*set/ weight** all but **ia** [36]

Matrix of values for the weights or connection strengths in the network. The row index corresponds to the receiving unit, and the column index corresponds to the sending unit.

set/ config/ all [34]

Header for variables that are relevant to the network configuration. Generally these are specified in the *.net* file.

*set/ config/ bepsilon** **bp** [143]

Vector of modifiability parameters, associated with bias terms in **bp**. See note under *set/ config/ epsilon* below.

*set/ config/ epsilon** **bp** [143]

Matrix of modifiability parameters associated with weights in **bp**. Note that nonzero modifiability parameters are set to the new value of *lrate* when *lrate* is changed.

set/ config/ ninputs all but **ia** [36, 107]

The number of input units. In **iac**, **cs** and **aa**, this is used only when reading patterns from a file, and is generally set equal to *nunits*. In **pa**, **bp**, and **cl**, it is used in specifying the network configuration as well as the elements in each input pattern.

set/ config/ noutputs **pa, bp, cl** [106]

The number of output units.

set/ config/ nunits all but **ia** [36, 107]

The total number of units in the network.

*set/ config/ uname** **cs, iac** [36]

Vector of unit names. Installed when the *get/ unames* command is used to specify a set of unit names.

set/ env/ all [34]

Header for variables associated with the learning environment.

set/ env/ ipattern **iac, cs, aa, cl** [36]
pa [106]
bp [143]

Matrix of input pattern element values. The row index corresponds to the pattern number, the column index corresponds to the element number within the pattern.

set/ env/ maxpatterns all but **ia** [37]

The maximum number of patterns that can be read in. This is generally automatically incremented when necessary, but if a large number of patterns are to be read in reading occurs faster if *maxpatterns* is preset to a number greater than or equal to the number of patterns in the pattern file.

set/ env/ npatterns all but **ia** [37, 106]

The number of patterns in the pattern list.

set/ env/ pname all but **ia** [36, 106]

Vector of names associated with the patterns on the pattern list.

set/ env/ tpattern **pa** [106]
bp [143]

Matrix of target pattern values. The row index corresponds to the pattern number, and the column index corresponds to the element number within the pattern.

set/ mode/ all [34]

Header for variables associated with various mode switches. These are generally used to select among variants of the same general model.

set/ mode/ boltzmann **cs** [57]

When nonzero, causes the **cs** program to simulate a Boltzmann machine.

set/ mode/ bsb **aa** [172]

When nonzero, causes the **aa** program to simulate the brain-state-in-a-box model.

set/ mode/ cascade **bp** [143]

When nonzero, causes the **bp** program to accumulate net inputs gradually rather than in a single step.

set/ mode/ clamp **cs** [57]

Determines whether external inputs are treated as specifying values that the specified units should be clamped with, or as specifying inputs to the units which are scaled by *estr* and then added into each unit's net input.

set/ mode/ comprp **ia** [226]

Determines whether response probabilities are calculated in the **ia** program. If *comprp* is set to 2, response probabilities are computed for letter and word units. If set to 1, they are computed for letter units but not for words. If set to 0, no response probabilities are computed.

set/ mode/ cs **pa** [106]

When nonzero, causes **pa** to set the activations of output units according to the continuous values returned by the logistic function.

set/ mode/ follow **bp** [143]

When nonzero, causes the **bp** program to follow the direction of the gradient from one weight update to the next by computing the correlation of the weight error derivative vectors used in each successive weight update.

set/ mode/ gb **iac** [37]

When nonzero, causes updating of activations to be done according to Grossberg's update rule in the **iac** program.

set/ mode/ harmony **cs** [58]

When nonzero, causes **cs** to simulate Smolensky's harmonium model.

set/ mode/ hebb **aa** [107]
pa [173]

When nonzero, causes **pa** and **aa** to use the Hebb rule rather than the delta rule in learning.

set/ mode/ lgrain **bp** [144]

Determines the grain of learning in **bp**. When *lgrain* is set to *pattern*, weights are incremented after each pattern is processed. When *lgrain* is set to *epoch*, weight error derivatives are accumulated over an entire processing epoch and then the weights are incremented only at the end of the epoch.

set/ mode/ linear **aa** [107]
pa [173]

When nonzero, causes **pa** and **aa** to use a linear activation rule instead of the default activation rules for these programs. See the relevant chapter for the exact update formulas used in each case.

set/ mode/ lt **pa** [107]

When nonzero, causes output units in the **pa** program to act like linear threshold units.

set/ mode/ selfconnect **aa** [173]

When nonzero, allows the **aa** program to adjust the strength of the connection to each unit from itself. Otherwise these connections stay fixed at 0.

set/ param/ all [34]

Header for parameter variables. These generally adjust *model* parameters, such as the rate of learning, rate of activation, etc.

set/ param/ alpha **iac** [37]

Scales the strength of excitatory input to each unit.

set/ param/ alpha/ **ia** [226]

Scales the strength of the excitatory input from the specified level to the specified level.

set/ param/ beta/ **ia** [226]

Scales the strength of the decay at the specified level.

set/ param/ crate **bp** [144]

Determines the rate of growth of the net input to each unit in cascade mode.

set/ param/ decay **aa, iac** [37]

Decay rate parameter.

set/ param/ estr **cs, iac** [37]
aa [173]

Scales the strength of external input (input from outside the network).

set/ param/ estr/ **ia** [227]

For each letter position, scales the strength of input from feature to letter units.

set/ param/ fgain **ia** [227]

Scales resting activation levels of word units based on word frequency.

set/ param/ fdprob/ **ia** [227]

Sets the probability that features are detected in each display field specified by the *trial* command.

set/ param/ gamma **iac** [37]

Scales the strength of inhibitory input to each unit.

set/ param/ gamma/ **ia** [227]

Scales the strength of inhibitory input from the specified level to the specified level.

set/ param/ istr **cs** [58]
aa [173]

Scales the strength of internal input (input from other units in the network).

set/ param/ kappa **cs** [58]

Threshold parameter used in *harmony* mode.

set/ param/ lrate **pa** [107], **bp** [144]
aa [173], **cl** [197]

The learning rate parameter.

set/ param/ max **iac** [37]

Maximum activation allowed for units.

set/ param/ max/ **ia** [227]

Maximum activation allowed for units at the specified level.

set/ param/ min **iac** [37]
Minimum activation allowed for units.

set/ param/ min/ **ia** [227]

Minimum activation allowed for units at the specified level.

set/ param/ momentum **bp** [144]

Fraction of previous weight increment incorporated in each new weight increment on each weight adjustment.

set/ param/ mu **bp** [144]

Fraction of the unit's prior activation that is added in when the activation of the unit is set on the basis of prior activations, as indicated by negative elements in the input pattern. See pages 156-157 for further explanation.

set/ param/ noise **pa** [107]

Amount of random noise added to each input and target value during learning.

set/ param/ orate **ia** [227]

Rate of integration of time-averaged activation for computing response strengths.

set/ param/ oscale/ **ia** [227]

Scale factor used in relating time-averaged activations to response strengths.

set/ param/ pflip **aa** [173]

Probability of flipping the sign of each input pattern element during learning.

set/ param/ rest **iac** [37]

Resting activation level of units.

set/ param/ rest/ **ia** [227]

Resting activation of units at the specified level.

set/ param/ temp **pa** [107]

Temperature parameter used in scaling net inputs. Not subject to annealing in **pa**.

set/ param/ thresh/ **ia** [227]

Threshold that units on the specified level must exceed before they generate output to units on the specified receiving level.

set/ param/ tmax **bp** [144]

Actual target activation used for target elements set to 1.0 in the pattern list. Actual target for elements set to 0 in the pattern list is 1 − *tmax*.

set/ param/ wrange **bp** [144]

Range of variability used when initializing random weights.

set/ state/ all [34]

Header for variables associated with the current state of the network.

*set/ state/ activation** **aa, cs, cl, iac** [37]
bp [144]

Vector of unit activations.

*set/ state/ bed** **bp** [144]

Vector of bias error derivatives.

set/ state/ cpname all but **ia** [38, 107]

The name of the pattern being tested or just tested.

set/ state/ cuname **cs** [58]

The name of the unit just updated.

set/ state/ cycleno **aa, bp, cs, ia, iac** [37]

The current cycle number. (In **bp**, relevant only in *cascade* mode).

*set/ state/ dbias** **bp** [144]

Vector of most recent delta biases, that is, of changes made to the bias values.

*set/ state/ delta** **bp** [144]

Vector of delta terms associated with each unit.

*set/ state/ dweight** **bp** [145]

Matrix of most recent delta weights, that is, of changes made to the weights. Indexes as with weights.

set/ state/ epochno **pa, bp, aa, cl** [107]

The current epoch number.

*set/ state/ error** **pa** [107]
bp [145]
aa [173]

Vector or error values associated with units. These have a slightly different meaning in each of the three cases.

*set/ state/ excitation** **iac** [37]

Vector of excitatory inputs to each unit.

*set/ state/ extinput** **cs, iac** [38]
aa [173]

Vector of external inputs to each unit from outside the network.

set/ state/ gcor **bp** [145]

Correlation of the weight error gradient calculated for the most recent weight update with the gradient calculated for the preceding update.

set/ state/ goodness **cs** [58]

Degree of constraint satisfaction due to the current state of the network.

*set/ state/ inhibition** **iac** [38]

Vector or inhibitory inputs to each unit in the network.

*set/ state/ input** **pa** [107]

Vector of activations of input units.

*set/ state/ intinput** **aa, cs** [173]

Vector of internal inputs to each unit from other units in the network.

set/ state/ ndp **pa** [107]
aa [173]

Normalized dot product of the current activation pattern with the target pattern.

*set/ state/ netinput** **cs, cl, iac** [38]
pa [107]
bp [145]

Vector of net inputs to each unit, including external inputs, internal inputs, and bias terms if any.

set/ state/ nvl **aa** [108]
pa [173]

Normalized vector length of the output activation vector.

*set/ state/ output** **pa** [108]

Vector of activations of output units. First element indexes the first output unit.

set/ state/ patno **iac, cs, cl, aa** [38]
pa, bp [108]

Number of the current pattern being tested.

*set/ state/ prioract** **aa** [174]

Vector of previous activation values of units in the network.

set/ state/ pss **aa, bp, pa** [108]

Sum of squared errors for the pattern most recently tested.

*set/ state/ target** **pa** [108]
bp [145]

Vector of target values for output units.

set/ state/ temperature **cs** [58]

Current temperature used to scale net inputs. The parameter is adjusted during simulated annealing.

set/ state/ tss **aa, bp, pa** [108]

Total sum of squares over all patterns tested in the current or just-completed epoch.

set/ state/ unitno **cs** [58]

The unit number of the unit most recently updated.

set/ state/ updateno **cs** [58]

The update number within the present processing cycle; equivalent to the number of updates done since the beginning of the cycle.

set/ state/ vcor **pa** [108]
aa [174]

The vector correlation of the pattern of activation with the external input (in **aa**) or the target (in **pa**).

*set/ state/ wed** **bp** [145]

Matrix of weight error derivatives. Indexes as with weights.

APPENDIX **C**

File Formats for Network Weight, Template, Look, and Pattern Files

This appendix describes the formats used in network, weight, template, look, and pattern files. These files are generally referred to as *.net*, *.wts*, *.tem*, *.loo*, and *.pat* files, respectively.

NETWORK FILES

The *.net* file is used to specify the network architecture for the following programs: **iac, cs, pa**, and **bp**. The **ia** program has a fixed network, and the **aa** and **cl** programs have networks of variable size but with a totally predictable architecture, so a *.net* file is not necessary.

The *.net* file consists of several sections, some of which may be optional. These sections are definitions, network, constraints, biases (for **cs**, **pa**, and **bp**), and sigmas (used in *harmony* mode in **cs**). Each section plays a different role in defining the network. The *definitions* section is used to specify the number of units in the network and also specifies other crucial variables. The *network* section is used to specify the pattern of connections among the units. Connections are of various types, each designated by a single letter. These connection types are defined in the *constraints* section, so called because the different types of connections are specified by constraints on the values they can take. The *biases* and *sigmas* sections specify characteristics of the *biases* and *sigmas* (if any) associated with the units in the network. Like connections, *biases* and *sigmas* can be of various types, each designated by a single letter defined in the constraints section. The definitions section of the *.net* file must come first, and is followed by the constraints section, if any (some connection types are predefined). Then comes the network section, followed by the biases and sigmas sections, if required.

The Sections of the Network File

Each section of a *.net* file begins with the name of the section (all lowercase), followed by a colon (e.g., *definitions:*) on a line by itself. The section ends with a line containing *end.* The following paragraphs describe the details of the information required in each section and the format of that information.

Definitions. The definitions section is used to specify basic parameters of the network architecture. For **iac** and **cs**, the only parameter that must be defined is *nunits.* If you wish to read patterns from a file, you will also have to define *ninputs*; this can be done in the definitions section, in the *.str* file, or by hand anytime before the *get/ patterns* command is issued. For **pa** and **bp**, *nunits, ninputs*, and *noutputs* must be defined. Other variables may also be initialized in the definitions, although they can also be set in the *.str* file.

Definitions are given by placing the name of the variable and the value you wish to assign to it on a line by itself. Thus, the definitions section of the *jets.net* file used with the **iac** program looks like this:

```
definitions:
nunits 68
end
```

Constraints. The constraints section of the file is used to define the meanings of the characters that are used in the network part of the file to designate the weight types. Each constraint definition is given on a separate line, consisting of a lowercase letter called the *constraint character*, followed by a list of constraint attributes. Constraint attributes can be:

- *A floating-point number.* The initial value assigned to each weight designated with the constraint character. If no value is given, the initial value will be 0, unless otherwise specified by another attribute. For unmodifiable weights, the weight will remain at this value.

- *Positive.* The weight is constrained to have a nonnegative value. This constraint is imposed after every weight adjustment; weights with this constraint that go below 0 are reset to 0.

- *Negative.* The weight is constrained to have a nonpositive value. This constraint is imposed after every weight adjustment; weights with this constraint that go above 0 are reset to 0.

- *Random.* The weight is initialized to a random value in a range given by the parameter *wrange.* Positive random weights vary between 0 and *wrange*; negative random weights vary between $-wrange$ and 0; otherwise the weight will vary between $+wrange/2$ and $-wrange/2$.

- *Linked.* The weight is constrained to have the same value as all other weights designated with this constraint character. This constraint is imposed at initialization and during each change to the weights. All the weights that are linked together are adjusted by an amount equal to the sum of the adjustments that would be made to each.

Several constraint letters are predefined. These are:

r Stands for random. The connection is set to a random value between $+wrange/2$ and $-wrange/2$.

p Stands for positive random. The connection is set to a random positive value between $+wrange$ and 0.

n Stands for negative random. The connection is set to a random negative value between $-wrange$ and 0.

. (The single character "."). This is the only character that is not an actual letter that can occur in the network specification itself. It specifies that the connection should be initialized to 0 and be treated as unmodifiable—that is, not subject to change through learning.

Note that positive, negative, and link constraints are only enforced in the **bp** program. In **pa** they are not used.

The constraints section of the *jets.net* file set up for use with **iac** looks like the following:

```
constraints:
u 1.0
d 1.0
v -1.0
h -1.0
end
```

Once this has been interpreted, it means that connections specified in the network section with the letters *u* or *d* will be assigned a weight of 1.0, and connections specified with letters *v* or *h* will be assigned a weight of −1.0. Though all the weights labeled *h* or *v* are assigned the same value according to this specification, it would be a trivial matter to dissociate the two by giving v, say, a value of -2.0; this is in fact what we did in the *jets.net* file that is set up for use with the **cs** program.

Network. The network section of the file specifies which of the defined constraint characters applies to each of the connections in the network. This part of the file can come in either of two formats. The more elementary format consists of a full matrix of *nunits* rows of characters, each *nunits* long and containing no tabs or spaces. This is the format used in the *jets.net* file. In this format, the entry in a particular row-column location specifies the connection to the unit whose index is the row number from the unit whose index is the column number. Thus in the *jets.net* file, the connection in row 0 column 1 specifies a connection to the Jets unit (unit 0) from the Sharks unit (unit 1). This connection is marked as a v, which has been defined to have the value -1.0. This connection will therefore be assigned the value -1.0.

Note that weight values are assigned according to the network specification file when the *.net* file is processed. Random weight values are reassigned in programs that learn (**pa, bp, aa,** and **cl**) when a *reset* or *newstart* command is executed. Note that the weight values assigned in the network specification file can be overridden, either by setting individual weights using the *set/ weights* command or by reading in a file of weights using the *get/ weights* command.

The network specification for the *jets.net* file is rather large, so we will use a simpler example instead: the *cube.net* file used with the **cs** program. This consists of a set of 16 rows of 16 dots:

```
network:
................
................
................
................
................
................
................
................
................
................
................
................
................
................
................
................
end
```

The actual values of these weights are set by reading in the *cube.wts* file using the *get/ weights* command.

Letters in the network specification can be uppercase or lowercase. If the letter is lowercase, the corresponding weight is modifiable; if it is uppercase, it is fixed. Of course, no weights are modifiable in programs that do not learn; thus for **iac** and **cs**, the characters *v* and *V* are synonymous.

A more complicated format for the network specification file is also available. In this format, connections are specified in *blocks*. A block specification is simply a specification of a portion of the full conceptual matrix of *nunits* by *nunits* connections. The block specification specifies which subpart of the matrix the specification applies to, as well as the characteristics of the connections in the block. The subpart is specified by indicating which units receive the connections specified in the block and which units send these connections to these receivers.

A block specification begins on a line consisting of a % character. The block-specification line also contains four integers. These integers represent:

1. The index of the first receiving unit in the block.
2. The number of receiving units in the block.
3. The index of the first sending unit in the block.
4. The number of sending units in the block.

For example, in a pattern associator network, there will generally be some number of input units, each projecting to some number of output units. The following specification can be used to set up such a network, with 12 input units and 8 output units:

```
network:
% 12 8 0 12
rrrrrrrrrrrr
rrrrrrrrrrrr
rrrrrrrrrrrr
rrrrrrrrrrrr
rrrrrrrrrrrr
rrrrrrrrrrrr
rrrrrrrrrrrr
rrrrrrrrrrrr
end
```

The %-line specifies a block of connections coming into unit 12 and the next 7 units (for a total of 8 receiving units) from units 0 and the next 11 units (for a total of 12 sending units). The next eight rows, one for each receiving unit specified on the %-line, each consists of 12 *r*'s, one for each sending unit specified in the %-line. The *r*'s indicate that these weights should be initially random and modifiable.

Note that if all of the weights in a block are to be of the same type, you can specify this by putting the letter specifying the type immediately following the %, *with no intervening space*. Thus the preceding example could be given more succinctly as follows:

```
network:
%r 12 8 0 12
end
```

Note that a network can consist of either a single, complete matrix of connections or of one or more block specifications. However, there is an important restriction:

A particular receiving unit *can only be specified in a single block*.

Finally, note that when block specifications are used, weights are only allocated for the connections actually specified in the blocks, and files of weights are assumed to contain values only for the weights that have been allocated. A detailed specification of the format of such files is given below.

Biases. The biases section of the file (applicable only to **pa, cs,** and **bp**), like the network section, can be given in either of two formats. In the simpler case, it consists of a row of *nunits* characters indicating the characteristics of the bias terms for all of the units in the network. Alternatively, biases may be specified in blocks, analogous to those used in the network specification. Block specifications consist of a line beginning with a %, followed by two integers that indicate the first unit in the block and the number of units in the block. The following line then gives a row of characters indicating the specification for each bias in the block; or if each bias is to be specified with the same character, the character may be given directly after the %. For example, random biases for the eight output units in the pattern associator network mentioned earlier could be specified in either of three ways:

```
biases:
............rrrrrrrr
end

biases:
% 12 8
rrrrrrrr
end

biases:
%r 12 8
end
```

Note that when biases are in use, the programs allocate biases for all units in the network, even when only a subset of the biases are specified using block notation.

Sigmas. The sigmas section of the file is analogous to the biases section, though it is applicable to *harmony* mode in the **cs** program only. The values associated with the specification characters are taken to specify the value of the parameter *sigma* for units in a harmony network.

An Example Network File

Here we give a complete example of a *.net* file, taken from the *xor.net* file used with the **bp** program. It specifies a network with two input units, one output unit, and two hidden units, with initially random, modifiable connections from each input unit to each hidden unit and from each hidden unit to the output unit. It also specifies random, modifiable bias terms for the hidden units (units 2 and 3) and the output unit (unit 4).

```
definitions:
nunits 5
ninputs 2
noutputs 1
end
network:
%r 2 2 0 2
%r 4 1 2 2
end
biases:
%r 2 3
end
```

The first line in the network section specifies that the hidden units (units 2 and 3) receive connections from the input units (units 0 and 1) and that these connections are modifiable, with initially random connection strengths. The next line specifies that the output unit (unit 4) receives connections from the hidden units (2 and 3), which are also modifiable, with initially random values. The biases section indicates that the biases of units 2 through 4 are also modifiable and initially random.

WEIGHTS FILES

The *.wts* files can be read into or written out from all of the programs other than **ia**, which has no matrix of weights as such. The format of these files is governed by the network specification file, according to the following conventions.

The *.wts* file consists of a list of the weights in the network, followed by a list of the bias terms, if any. The list of weights can be thought of as consisting of a series of rows, one for each unit with incoming connections from other units. Rows are ordered by unit number, from first to last. Each row consists of one floating-point number for each connection to the receiving unit it applies to. These numbers specify the values of these connections, ordered by unit number, from first to last. Row elements must be separated from each other by any number of spaces, tabs, or newline characters. Thus, the order of entries is crucial, but spaces, tabs, and newlines[1] can be used freely for readability without affecting the way the file is interpreted. In fact, the *save/ weights* command places only a single weight on each line of the file it produces, even though the weights are conceptually grouped into rows. This is done to avoid the possibility of exceeding line-length limitations imposed by the file system on your computer.

In the simplest case, the *.net* file will specify a full matrix of *nunits* by *nunits* connections. In this case the list of weights in the *.wts* file will consist of *nunits* rows, one for each unit, and each row will contain *nunits* floating-point numbers, one for the connection of each unit to the row unit.

More complex cases arise when weights have been specified using block specifications. In this case, rows of weights are only given for units that actually receive connections from other units, and each row will only have entries for the connections specified in the block specification involved. Thus the *xor.wts* file, which is used in the XOR example in Chapter 5, contains only three rows of weights, one for each of the two hidden units and one for the output unit. Each row contains only two entries since each of these units receives a connection from only two other units.

Following the list of weights comes the list of biases, if any. Note that if a biases section is present in the *.net* file, a full list of *nunits* biases is expected in the *.wts* file, even if only a subset of the bias terms were actually specified in the *.net* file. Each bias is a floating-point number, separated from others by spaces, tabs, or newlines.

An Example Weights File

As an example of a *.wts* file, we give here the file *xor.wts*. It was created using the *save/ weights* command and is read into the **bp** program using the *get/ weights* command, after the *xor.net* file has been used to set up the network. Unspecified biases are simply stored as 0.00000. Here is the file:

[1] The *newline* character is the character in a file that separates lines from each other.

```
0.432171
0.448781
-0.038413
0.036489
0.272080
0.081714
0.000000
0.000000
-0.276589
-0.402498
0.279299
```

Note that the weights file contains no special indications about where the rows of weights end or even where the weights end and the biases begin. This is because the number of conceptual rows of weights, the number of weights per row, and the number of biases is determined strictly by the specifications contained in the *.net* file.

TEMPLATE FILES

The template file is used to specify the appearance of the display screen and the way in which various display objects, called *templates*, will appear on the screen. The file consists of an optional *layout*, followed by a series of *template specifications*. The layout is used to set up the background on which the various templates will be displayed and to specify where the templates will occur in the background. If the layout is omitted, template locations can be specified directly. We will first describe the format for the layout, then describe the template specifications.

The Layout

If there is a layout, it must occur at the very beginning of the template file. The first line of the file must be

```
define: layout
```

The first line can also contain two integers indicating the number of rows and columns to use on the screen. By default, the program assumes it can use 23 rows (lines) and 79 columns. This is one row and column less than the typical screen size, to make it easy to review screen dumps simply by using the *type* command in MS-DOS. Twenty-three lines of 79 characters will fit on the screen without scrolling off the top. To change this default, give the desired number of rows and columns after the word *layout*. For example, to increase the size of the display to 47 by 131, the first line would be

```
define: layout 47 131
```

Subsequent lines are taken as containing literal characters to plot directly to the screen and $'s, which specify where templates are to be inserted. No tabs are allowed in the layout; you must use spaces as separators, although it is perfectly acceptable for lines to be blank or less than 79 characters long. The layout thus gives an exact image of what the screen is to look like. The end of the layout is indicated by a line containing the single word *end*. The layout is printed to the screen starting on line 5, and may be up to $nr - 5$ lines long where nr is the number of rows. Five rows are reserved for the command line at the top of the screen and the help area just below it.

An Example Layout From a Template File

The layout for the cube example used with the **cs** program (in the file *cube.tem*) is specified as shown in Figure 1. The layout is read from top to bottom and, within each row, from left to right; the locations of the $'s are stored in the order encountered. Successive $'s in the layout are numbered starting from 0. Below we explain how the $'s are paired up with templates.

The Template Specifications

Each template specification consists of a number of entries, or *template specifiers*, separated by white space (spaces, tabs, or newlines). To illustrate, we will describe the template specification from the *cube.tem* file that is responsible for putting the value of the variable *cycleno* in the correct place on the screen. The specification is

```
cycleno variable 1 $ 2 cycleno 3 1
```

The entries correspond to the following specifiers. The specifiers are listed in the order encountered, with their values from the *cube.tem* example given in parentheses.

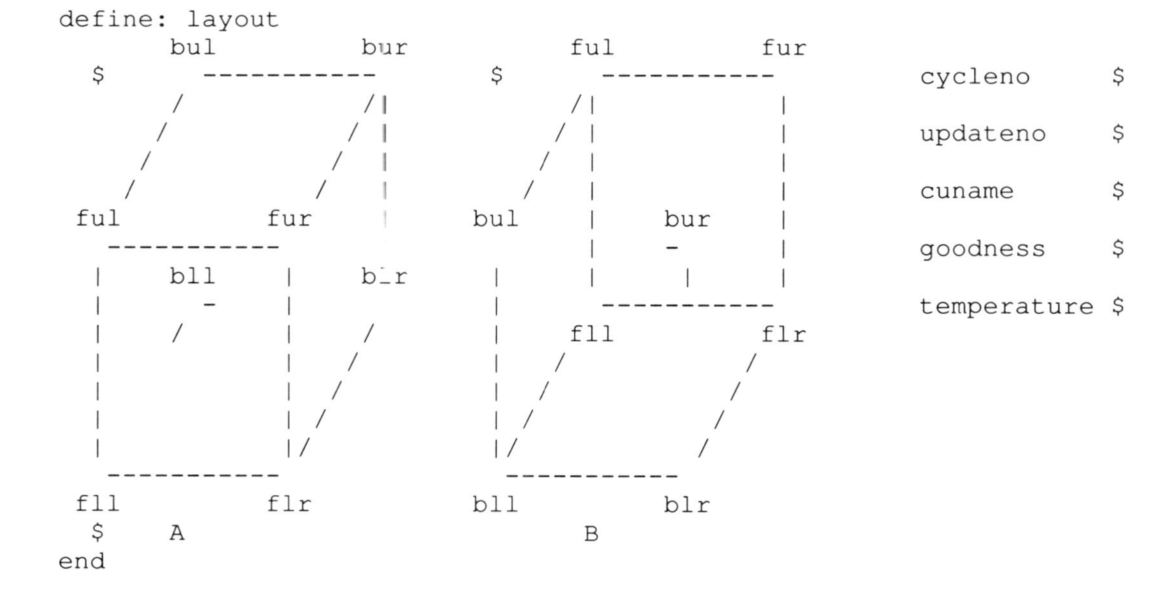

```
define: layout
       bul         bur          ful         fur
  $      -----------       $      -----------         cycleno     $
       /           /|           /|           |
      /           / |          / |           |        updateno    $
     /           /  |         /  |           |
    /           /   |        /   |           |        cuname      $
 ful         fur    |     bul    |    bur    |
   -----------                   |    -      |        goodness    $
  |    bll    |    blr     |     |     |     |
  |      -    |            |      -----------         temperature $
  |    /      |    /       |    fll         flr
  |           |   /        |   /           /
  |           |  /         |  /           /
  |           | /          | /           /
  |           |/           |/           /
   -----------              -----------
 fll         flr          bll         blr
  $    A                         B
end
```

FIGURE 1.

TNAME (*cycleno*)

The name for the template, or display object, itself. This name can be used as an argument to the *display* command, and will cause the template to be displayed on the screen. Thus to display this template you would enter

display cycleno

Note that a number of templates can have the same name. If such a name is given to the *display* command, all of the templates with that name will be displayed together.

TYPE (*variable*)

The template type of the template. There are a number of different template types, which will be described in the next section. The *variable* template type is used for most single-valued variables such as *cycleno*.

DLEVEL (1)

The display level of the template. The display level is an integer that determines whether the template will be displayed when the screen is updated by the program. There is a global *dlevel* variable and a *dlevel* value associated with each template. Templates are updated if their specific *dlevel* is less than or equal to the global *dlevel*. However, templates whose specific *dlevel* is 0 are treated specially. Generally, these are templates that display information that is not expected to change in the course of processing. These templates are only displayed if the screen has been cleared since the last time the screen was updated.

POS ($ 2)

This compound entry specifies the row and column location of the upper left-hand corner of the screen region in which the template is to be displayed. The "$ 2" notation is used to specify which $ in the layout should be taken as the one indicating the upper left corner of this template. Note that $'s are numbered starting with 0, in the order they are encountered in processing the layout, as specified above. An alternative notation is "$ *n*," where the *n* stands for *next* (note that *n* is used literally here). This notation is taken to mean the next $ after the one specified in the previous template specification. If no templates have already been specified, *n* is taken to mean 0. As a third alternative, the compound entry can be two integers. These are taken to indicate directly the row and column number of the upper left corner of the template. Note that the row number must be 5 or greater and must be less than the number of rows.

VNAME (*cycleno*, second occurrence)

The name of the variable as it is known in the program. Note that this is not the same as the name of the template, mentioned above;

two templates with different names can display the same variable. In general, any of the variables that can be accessed by the *set* and *exam* commands can also be accessed in templates.

SPACES (3)
: The number of spaces the program is to use in displaying this template.

SCALE (1)
: A floating-point number that specifies a scale factor that is multiplied with the value of the number to be displayed. This is particularly useful for floating-point variables that are to be displayed in small fields without decimal points.

Other specifiers are more specific to particular template types. These are defined in discussing each template type below.

The Template Types

Template specifications always begin with the TNAME, followed by the TYPE. The following list describes the different types that are available, together with a list of the specifiers that must come after the TYPE. Note that for each template, *all* of the specifiers must be present in the order indicated. Remember also that the POS specifier is a compound specifier, consisting of two entries separated by white space.

label (DLEVEL POS ORIENT SPACES)
: The TNAME of this template is taken as a literal label to be displayed either horizontally or vertically, as determined by the value of ORIENT (*h* or *v*). The label is truncated if it is longer than SPACES characters.

variable (DLEVEL POS VNAME SPACES SCALE)
: The single-valued variable VNAME is displayed. The variable may be an integer, a floating-point number, or a character string. Integers and floating-point numbers are multiplied by SCALE before the value is displayed. In either case, the value is truncated to an integer. Thus a variable whose internal value is 2.67 will display as a 2 if SCALE is 1, as 26 if SCALE is 10, and as 267 if SCALE is 3. Negative numbers are generally displayed in reverse video. Numeric values that would take up less than the available number of spaces are right-justified; this is typically appropriate for numbers. Values that would take up too much space are displayed according to the conventions described in Chapter 2. For character string variables (e.g., pattern names), if the value of the SPACES field is negative then the string is left-justified. Thus "−5" means that five spaces are allocated and that the string always begins at the

left end of the region. SCALE is not used for character strings, but a value must be specified anyway.

floatvar (DLEVEL POS VNAME SPACES SCALE)
The value of the scalar floating-point variable VNAME is displayed to four decimal places, in standard floating-point notation. The value is multiplied by SCALE before displaying and is right-justified within the SPACES indicated. If the value is too big, it is truncated on the right so that it fits the available SPACES.

vector (DLEVEL POS VNAME ORIENT SPACES SCALE START NUMBER)
NUMBER elements of the vector VNAME are displayed, starting with element START. The vector may consist of integers, floating-point numbers, or character strings. For example, a list of unit names can be displayed as a vector template. If ORIENT is *h*, the elements of the vector are displayed across the same line. If ORIENT is *v*, the elements are displayed starting in the same column, on successive lines. In both cases multiple characters making up the same element are displayed horizontally. For numeric variables, each element is multiplied by SCALE and then truncated to an integer, as described above for *variable* templates. For vectors of character strings, SCALE (which should have an integer value in this case) is interpreted as a specifier indicating the number of blanks to place at the beginning of the string. This allows the user to put blanks between the successive elements of a string vector when displayed horizontally.

label_array (DLEVEL POS VNAME ORIENT SPACES START NUMBER)
This template is similar to the preceding one, but it is used only for vectors of character strings or labels, and it allows strings to be displayed horizontally or vertically. The value of the ORIENT specifier applies to the orientation of the individual strings. If ORIENT is *h*, each individual string is shown horizontally, and successive strings are shown on successive lines. If ORIENT is *v*, each individual string is shown vertically, and successive strings are shown in successive columns.

look (DLEVEL POS VNAME SPACES SCALE SEPARATION LOOK-FILE)
This template type can be used with either vector or matrix variables. It consults the LOOKFILE (which must be present in the directory the program is being run in) for a specification of a rectangular grid of locations in which to display elements of the vector or matrix. Each element is multiplied by SCALE and then truncated to an integer, as described above for *variable* templates. SPACES character positions are allowed for each element; adjacent elements are SEPARATION spaces apart. Thus if SPACES is 2, SEPARATION is 4, and SCALE is 10, three adjacent elements with the values 1, 0.5, and 7 would appear as

```
10   5  70
```

label_look (DLEVEL POS VNAME ORIENT SPACES SEPARATION LOOKFILE)
This template type is similar to the previous one, but is used with arrays of strings or labels, such as unit names and pattern names. ORIENT is *h* or *v*, indicating whether individual strings should be displayed horizontally or vertically.

matrix (DLEVEL POS VNAME ORIENT SPACES SCALE ST_ROW N_ROWS ST_COL N_COLS)
A section of the matrix VNAME is displayed. The section begins in row ST_ROW and column ST_COL, and is N_ROWS by N_COLS in size. Note that when the matrix is a weight matrix, the row indexes correspond to the units receiving the connections and the column indexes correspond to the sending units. Thus to show the weights for connections to units 4 through 7 from units 0 through 3, ST_ROW would be 4, N_ROWS would be 4, ST_COL would be 0, and N_COLS would be 4. Also note that arrays of vectors are essentially matrices. Thus, for example, *ipattern* is a matrix variable, consisting of *npatterns* rows, each *ninputs* long. In this case, then, the row indexes correspond to patterns and the column indexes to elements of the patterns. The ORIENT field specifies the orientation of the rows of the matrix. Thus if ORIENT is *h*, the rows are displayed horizontally and the columns are displayed vertically. If ORIENT is *v*, the rows are displayed vertically and the columns displayed horizontally. This is equivalent to transposing the matrix.

An Example List of Template Specifications

As a full example, the template list from the file *cube.tem* is shown here:

```
cube1        look      1 $ 0 activation   1 10 1 cube1.loo
cube2        look      1 $ 1 activation   1 10 1 cube2.loo
cycleno      variable  1 $ 2 cycleno      7  1
updateno     variable  1 $ 3 updateno     7  1
uname        variable  1 $ 4 cuname      -7  1
goodness     floatvar  1 $ 5 goodness     7  1
temperature  floatvar  1 $ 6 temperature  7  1
weight       matrix    5 $ 0 weight   h   4 10 0 16 0 16
weight       vector    5 $ 2 uname    v   6  1 0 16
weight       vector    5 $ 7 uname    h   4  1 0 16
```

The POS specification consists of a $ followed by a number. Each template is, therefore, displayed at the location of the $ whose index is given by the second part of the POS specification. Most of the different template types are illustrated here, as well as several useful tricks. For example, there are three templates called *weight*. When the user enters *display/ weight*, all three are displayed simultaneously. In this way, it is possible to put together "macrotemplates" from several ordinary templates.[2] Another trick is the use of the SCALE specifier with vectors of character strings. This occurs in the last entry, which specifies a horizontal array of unit names. The SPACES field indicates that 4 characters are allocated to each unit name, but the SCALE field indicates that each unit name is to be left-padded by a blank. Thus the four spaces are occupied by a blank, followed by the first three characters of the unit name.

LOOK FILES

The look files specify where in a rectangular array the elements of a vector or matrix are to be displayed. The format of the look file is as follows: The first line contains two integers, indicating the number of lines of entries and the number of entries per line. Below this, there are as many lines as specified in the first argument, each containing as many entries as indicated by the second argument.

When the object being displayed is a vector variable, the entries may be integers, in which case they are taken to be the index specifying the element of the vector that is to be printed at the corresponding location in the display. Alternatively, entries may simply be single dots ("."), which are taken to indicate that nothing is to be displayed at this location.

An Example Look File

The file *cube1.loo* is shown here as an example. This look file is used by the first template that was specified in the *cube.tem* file to indicate where, with respect to this template's upper left-hand corner, the 0th through 7th elements of the activation vector are to be displayed. Entries in the look

[2] A variant of this involves naming a group of templates with names that begin the same, but end differently. This way if the user enters the initial part that is shared, all of the variants are displayed, but if the user enters enough of the string to specify the template uniquely only the unique template that matches the full string entered is displayed. For example, the environment, including the set of *pnames* and *ipatterns* (and possibly *tpatterns*) is often specified by templates called *env.pname*, *env.ipat*, and *env.tpat*.

file itself must be separated by white space (spaces, tabs, or newlines), but these do not affect the actual appearance of the display. The spacing of the display itself is controlled in part by the SEPARATION variable associated with the template; this variable indicates the number of successive spaces between look elements along the same line.

```
15 19
. . . . . . 0 . . . . . . . . . . . 1
. . . . . . . . . . . . . . . . . . .
. . . . . . . . . . . . . . . . . . .
. . . . . . . . . . . . . . . . . . .
. . . . . . . . . . . . . . . . . . .
. . . . . . . . . . . . . . . . . . .
2 . . . . . . . . . . . 3 . . . . . .
. . . . . . . . . . . . . . . . . . .
. . . . . . 4 . . . . . . . . . . . 5
. . . . . . . . . . . . . . . . . . .
. . . . . . . . . . . . . . . . . . .
. . . . . . . . . . . . . . . . . . .
. . . . . . . . . . . . . . . . . . .
. . . . . . . . . . . . . . . . . . .
6 . . . . . . . . . . . 7 . . . . . .
```

An Example Look File for a Matrix Variable

When using look files with matrices (such as weight matrices), each non-dot entry specifies the row and column in the underlying internal matrix of the element to be displayed at that location. Row and column numbers are separated by a comma. As an example, the following look file is used to specify a layout for the weights coming into each of two hidden units from each of sixteen input units. This is the file *16wei.loo*, used with the **cl** program. The file lays the weights out in two square arrays, one for the connections coming into one of the units and one for the connections coming into the other. The column of dots down the middle separates the two arrays.

```
4 9
 16,0  16,1  16,2  16,3 .  17,0  17,1  17,2  17,3
 16,4  16,5  16,6  16,7 .  17,4  17,5  17,6  17,7
 16,8  16,9 16,10 16,11 .  17,8  17,9 17,10 17,11
16,12 16,13 16,14 16,15 . 17,12 17,13 17,14 17,15
```

PATTERN FILES

Pattern files are of two kinds. One kind contains a list of input patterns; the other contains a list of input-target pattern pairs. The first kind is used with **iac**, **cs**, **aa**, and **cl**. The second kind is used with **pa** and **bp**. We describe the input pattern type first.

Files for Lists of Input Patterns

Files containing lists of input patterns consist of a sequence of patterns. Each pattern consists of a name followed by *ninputs* entries specifying the values of the elements of the pattern. The entries are separated by white space (spaces, tabs, or newlines).

Each name is a string of characters beginning with a letter; pattern names should not begin with any of the digits. By convention we use the letter *p* as the first letter of the name of any pattern which for mnemonic reasons would begin with a digit.

The entries specifying the values of elements of the pattern are treated as floating-point numbers. The following special single-character entries are recognized:

. (dot) is assigned the value 0.0.
+ (plus) is assigned the value +1.0.
− (minus) is assigned the value −1.0.

Files for Lists of Input-Target Pairs

These files contain a sequence of pattern pairs. Each pair consists of a name, *ninputs* entries specifying the elements of the input pattern, and *noutputs* entries specifying the elements of the output pattern. Entries are separated by white space as in input pattern files. There is no special separation between input and target patterns. Both input and target entries are treated as floating-point numbers, with the same special characters recognized as in input pattern files.

In **bp**, negative entries have special meanings that are different for input pattern elements and target pattern elements. A negative input pattern element is interpreted as an instruction to set the activation of the corresponding input unit to *mu* times its previous activation plus the activation of the unit whose index follows the minus sign. A negative target pattern element is interpreted as an instruction to ignore the output generated by the corresponding output unit. The error for this unit is set to 0.

Finally, the reader should note that in **bp**, the parameter *tmax* is used to modify target elements specified with entries of 1 or 0. If the entry is 1, the target is set to *tmax*, and if the entry is 0, the target is set to $1 - tmax$. By default *tmax* is set to 1.0, but sometimes it is useful to use a value like 0.9. In this case, if the entry is 1, the target is set to 0.9, and if the entry is 0, the target is set to 0.1.

APPENDIX D

Plot and Colex: Utility Programs for Making Graphs

The programs described in this appendix are intended to allow the user to make simple two-dimensional graphs based on information logged while running one of the simulation models in this package.

We mention first the format of the log files. We then describe how to extract information from them using **colex** and how to make a graph using **plot**.

LOG FILE CONVENTIONS

The simulation programs log information according to specifications contained in the template file. Logging occurs only if a log file has been opened using the *log* command. If it has, then every time the screen is updated, all templates that are displayed are candidates for logging. A template is displayed if its *dlevel* is positive and is less than or equal to the value of the global *dlevel* variable.[1] A displayed template is logged if the template's local *dlevel* is less than or equal to the global *slevel* variable.

On each update, all of the logged variables are stored on a single line, separated from each other by spaces. Templates are considered for display and for logging in the order encountered in the template file. Within a template, values of vector templates are displayed in order; values of matrix templates are displayed in order as well, with the column index changing most rapidly. Values of look templates are displayed in the order they are specified in the look file, which is read like a matrix, with the column index

1 Note that templates with a *dlevel* of 0 are only updated when the display is clear, and are never logged.

changing most rapidly. In general then, for look templates, the order in which values are displayed and logged will not necessarily correspond to the order in which they are stored internally in the program.

As an example, the template list from the *jets.tem* file used with the **iac** program looks like this:

```
cycleno     label       0 5 70            h   5
cycleno     variable    1 5 75 cycleno    4   1
unitnames   label_look  0 5  3 uname      h   7 14 jets.loo
extinput    look        2 5  0 extinput   2 100 14 jets.loo
activation  look        1 5 10 activation 3 100 14 jets.loo
```

The third entry in each template specification is the *dlevel*. If the global *dlevel* is 2 and the global *slevel* is 1, then the *cycleno*, *extinput*, and *activation* templates will be displayed each time the display is updated, but only the *cycleno* and the *activation* will be logged.

Suppose that the **iac** program is called with this template and that the *dlevel* is set to 2, the *slevel* is set to 1, and the log command has been used to open the log file *jets.log* after the display is first initialized. Suppose further that the *stepsize* is set to *cycle*, so that the display is updated at the end of each cycle of training. Then when the program is given the *cycle* command, it will store a line for each cycle. Each line begins with the cycle number and then has one entry for the activation of each unit that is shown in the display. Each line is, therefore, rather long, and particular elements of the line must be extracted for further perusal or plotting.

Since the activations are displayed according to the specification stored in the *jets.loo* file, the order of logging of activations is given by the order in which units are encountered in the look file, which is read across each row. As examples, entry 1 (the one after the cycle number) corresponds to *Jets*, entry 2 corresponds to *Art*, entry 6 corresponds to *Sharks*, entry 15 corresponds to *in20s*, and entry 31 corresponds to the name unit for Ken.

EXTRACTING COLUMNS TO PLOT USING COLEX

The **colex** command is used for extracting columns from log files for plotting. Note that **colex** is a separate program, run under the command interpreter, rather than a command that is executed by the simulation programs themselves. Typically it would be used after a simulation session to plot the results of a simulation run.

The **colex** program is called in the following way:

colex infile outfile [abutfile] cols

where *infile* is the name of the input file from which you wish to extract columns, *outfile* is the name of the output file you wish to put the extracted columns in, *abutfile* (optional) is the name of a file containing lines you wish to abut the new output with,[2] and *cols* is a list of column numbers *counting from 0* to extract from the input file and place in the output file.

Let's say in our **iac** example that we have run 100 cycles of processing, logging the results as specified above. Now let's suppose we would like to see how the activation of a few of the units changed over processing cycles, say, the *Sharks* unit, the *in20s* unit, and the name unit for Ken. Before we can plot these we must extract them from the log file, and we will use **colex** for this purpose. As noted above, the *cycleno* is in column 0, and the activations of the *Sharks*, *in20s*, and *Ken* units are in columns 6, 15, and 31, respectively. Thus the command to pull these columns out of the file *jets.log* and store then in *jets.clx* would be

colex jets.log jets.clx 0 6 15 31

The optional *abutfile* argument to **colex** allows the user to specify a file containing lines to which the results of the column extraction operation are to be abutted, or appended to. As an example of the use of the abut facility, suppose that we ran the **iac** program twice, each time with a different value of the parameter *gamma*, and we wanted to see how this affected the activation of the unit for *in20s*; and suppose we had logged the runs in *run1.log* and *run2.log*. To plot the activation curves from the two runs on the same graph we would want to get them into the same file, say, *both.clx*, containing lines with the cycle number and the activation of the *in20s* unit from each of the two runs. We can do this with *colex* as follows:

colex run1.log run1.tmp 0 15
colex run2.log both.clx run1.tmp 15

MAKING GRAPHS WITH PLOT

The **plot** program is used to make graphs from files containing lines, each containing one x-value and one or more y-values to plot against the single x-value. The x-value is always the first one in the line. The file *jets.clx* is an example in which there is one x-value and three y-values per line.

The plots are displayed in an array of 23 lines, each containing 79 characters followed by a newline character. The output of the program is normally displayed on the screen, but it can be dumped to a file if an optional output filename is supplied.

---

2 Note the square brackets here are used to indicate that the *abutfile* argument is optional and they should not be typed when running the program.

The **plot** program is called with the following command format:

plot formatfile datafile [outputfile]

Here *formatfile* specifies a file containing plot format instructions, the *datafile* specifies a file containing the lines of xy-values, and the optional *outputfile* specifies a file for storage of the output.

The format file consists of a number of format specifications. Each format specification consists of a line beginning with one of the following characters, each of which has the meaning indicated:

t Following argument specifies a title for the plot.

l Following argument specifies a label to associate with a column of y-values. There can be as many label specifications as there are y-values on each line. They are assigned to columns of y-values in the order encountered.

s Following single character specifies the symbol to be plotted at each xy-location. If no symbol specification is given, the program uses the letter *a* for the first y-value and uses successive letters of the alphabet for successive y-values. Where more than one xy-pair would fall on the same column-line position, the program plots a digit indicating the number of such collisions or "*" if there are 10 or more. If a symbol is specified, all of these features are overridden and the symbol is used for plotting every point.

x The rest of the line contains a specification of a parameter of the x-dimension of the plot.

y The rest of the line contains a specification of a parameter of the y-dimension of the plot.

The dimension specifications are as follows:

M Following argument specifies the maximum value for the dimension.

m Following argument specifies the minimum value for the dimension.

t Following argument specifies a title for the corresponding axis.

l Indicates that natural logarithms should be taken of the values on this dimension.

As an example, the following format specification is the one used to produce the graph shown in Figure 1:

```
t Activations_for_Ken
l Sharks
l in20s
l Name
x M 110
x m 0
y M 1.0
y m -.2
x t Cycles
y t Activation
```

If this specification is stored in the file *jets.fmt*, the graph shown in the figure can be generated using the following command:

plot jets.fmt jets.clx

To store it in the file *jets.plo*, the command would be

plot jets.fmt jets.clx jets.plo

Some discussion of the appearance of the figure is in order. The graph has very low resolution both horizontally and vertically; it uses only 20 rows and 60 columns for the actual graph itself. Consequently, there

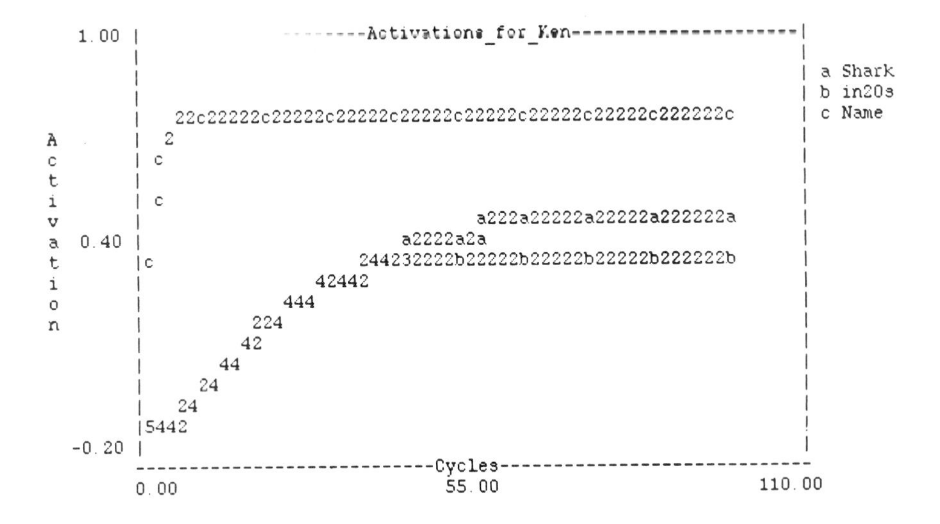

FIGURE 1. The graph showing the activation of the *Sharks*, *in20s*, and *Ken* units, from the file *jets.clx*, based on a run of the **iac** program with external input supplied to the *Ken* unit.

appears to be a succession of steps in some of the curves, though in fact the underlying data are much smoother. Note also that there are many 2s and 4s; these occur whenever more than one point is to be plotted in the same row-column location of the 20×60 array.

In spite of the low resolution, the figure is somewhat revealing; it illustrates the fact that the name unit for Ken reaches asymptote very early in processing (this is due to the fact that it receives external input in this run). The other two units rise rather gradually to their asymptotic activation values of 0.50 and 0.38.

APPENDIX **E**

Answers to Questions in the Exercises

CHAPTER 2

Ex. 2.1. Retrieval and Generalization

Q.2.1.1. Some of Ken's properties are more strongly activated than others because several instance units other than the instance unit for Ken are partially active. These instance units send activation to their property units. In particular, the units for Nick and Neal are both reasonably active. They are active because they share *Single*, *Sharks*, and *HS* with Ken; they in turn support these properties of Ken, reinforcing their activation relative to others.

Q.2.1.2. The instance units for some of the individuals eventually receive considerable excitation because they are supported by several of the active property units. In contrast, the name units for these same individuals only receive excitation from one source—that is, the corresponding instance unit. The inhibition coming from Ken's name unit is stronger than the excitation any of the other name units is receiving.

Q.2.1.3. All of the active instance units have three properties in common with Ken. However, the six active instance nodes differ in the extent to which their properties are shared by the other active instance units. Nick and Neal share *Sharks*, *HS*, and *Single* with

Ken, whereas Earl and Rick share *Sharks*, *HS*, and *Burglar*, and Pete and Fred share *in20s*, *HS*, and *Single*. All the properties Nick and Neal share with Ken are also shared with two of the others, whereas Earl and Rick are the only individuals who share *Burglar* with Ken, and Pete and Fred are the only ones who share *in20s* with Ken. Because the properties that most of this group share receive more top-down support, the individuals who share them get more bottom-up activation. This effect is known as the "gang effect" (McClelland & Rumelhart, 1981) and is discussed again in Chapter 7.

Q.2.1.4. This is a result of hysteresis. The *in20s* unit became active as a result of the activation of the instance unit for Ken. At the same time other property units became active and these eventually began to activate other instance units, which in turn sent excitation to the *in30s* unit. By this point, however, the *in20s* unit was already active, so it continues to send inhibition to the *in30s* unit, keeping it from becoming as active as it would otherwise tend to become.

Q.2.1.5. Among the instance units, only four "competitors" of Ken are active; these are the ones that match one of the two properties of the probe and that share the *HS* and *Single* properties with Ken. Even though Earl and Rick share three properties with Ken, they are not active in this case. Among the property units, the properties shared by all the active instance units are more active than before, whereas the *Burglar* unit is less active; the other occupation units are now above threshold.

Q.2.1.6. In part because the probe directly excites them and in part because they agree on two out of the three properties they share with Ken, the instance units for Pete, Fred, Nick, and Neal are all quite active in this case. Since two of these support *Pusher* and two support *Bookie*, these two units receive more activation than in the previous run. Also, since in the first run the instance unit for Ken was the only instance unit active for some time, the *Burglar* unit was able to establish itself more strongly in the early cycles, thereby helping to keep the other property units from exceeding threshold.

Q.2.1.7. The noisy version of Ken "works" in this case because the probe is still closer to Ken than to any other instance known to the system. In other cases, distorting one property does not work because sometimes it produces a probe that is an equally good match to two or more instances, and sometimes it even produces a probe that is a perfect match to a different individual from the one that matched the undistorted version of the probe.

Q.2.1.8. Success is due to the fact that several instance units that share several of Lance's properties also happen to share the same occupation. The model generalizes on the basis of similarity, and in this instance it happens to be right. Activation of the *Divorced* unit comes about because two of the instances that are similar to Lance are divorced, rather than married. Guilt by association.

Q.2.1.9. We leave this question to you.

Ex. 2.2. Effects of Changes in Parameter Values

Q.2.2.1. The effect of decreasing these four parameters is simply to slow processing by a factor of 2; it is as if time runs more slowly. One can easily show that the equilibrium activations of units should not be affected by proportional increases in all these parameters since they can be factored out of both the numerator and denominator of Equation 2 (p. 13). The effect of increasing these four parameters, however, is not so simple because it introduces abrupt transitions in the activations of units. Basically, what happens is that a large number of units (for example, at the instance level) suddenly receive enough bottom-up activation to become active. Since none were active before, they were not receiving any inhibition at this point. When they all suddenly become active, however, they all start sending each other inhibition. This causes them all to be shut off on the next cycle, even though all are still receiving bottom-up excitatory input. Since they are all off at this point there is no inhibition, so on the next cycle they all come on. This "ringing" effect eventually sets up an oscillatory pattern that destroys the information content of the pattern of activation.

Q.2.2.2. Increases in *gamma* limit activation of partially matching instances and also limit activation of partially supported properties. The result is a more precise and focused retrieval of the best match to the probe, with less influence from partial matches. In extreme cases this completely eliminates, for example, the "default assignment" process seen previously in the Lance example. Decreases in *gamma* allow more activation of partial matches; in the limit, these partial matches can swamp the best match, destroying the information content of the pattern of activation.

Ex. 2.3. Grossberg Variations

Q.2.3.1. The main point here is that in the standard update rule, the net input to the unit is computed regardless of the current activation of the unit; when the net input is positive it drives the activation of the unit up, and when it is negative it drives the activation of the unit down. In the Grossberg version, the net input is not computed as such; the magnitude of the upward force on the unit's activation (that is, the effect of the excitatory input) is scaled by the distance from the current activation to the maximum, while the effect of inhibitory input is scaled by the distance from the current activation to the minimum. In this regard it is useful to contrast the effect of equal excitatory and inhibitory input in the two cases. In the standard case, these cancel each other and so there is no change in the activation of the unit. In the Grossberg case, the effects are modulated by the current activation of the unit; the effects balance only when the activation is halfway between the maximum and minimum activations. In these simulations this means that the effects will balance at an activation of 0.4, halfway between 1.0 and −0.2. To compensate, one can increase *gamma*. For example, if *gamma* were set to 5 times *alpha*, then equal excitation and inhibition would have canceling effects when the unit's activation was exactly 0. Selecting such a value for *gamma* brings the two versions closer into line. However, this tends to result in lower overall activations, since as units become excited above 0, the power of their inhibitory inputs becomes much more potent.

One advantage of the Grossberg version is that it is in fact more stable; in general it appears that the standard version drifts away from the approximate equilibria we see at around 100 cycles. In contrast, the Grossberg version tends to find very stable fixed points.

CHAPTER 3

Ex. 3.1. The Necker Cube

Q.3.1.1. You should find that each of the two valid interpretations is reached about a third of the time, and local maxima are found on the other third. Of these, you should find some in which one side of cube A is on and the opposite side of cube B is on, and some in which one edge of one of the cubes is on and the other three edges of the other cube are on.

Q.3.1.2. Generally, what happens is that local minima are reached when the first few units updated are on edges of the two cubes that do not interact. For example, in one minimum, the units on the right surface of cube A are on, together with those on the left surface of cube B. This minimum tends to be reached if units on these two surfaces become active early on. Note that units on these surfaces do not directly inhibit each other, but do inhibit other units in the other's cube. These inhibitory influences prevent either cube from completing itself. In one particular run, of the first four units updated, two were on the left surface of cube A and two were on the right surface of cube B. Four of the next five updates were on these two surfaces. At this point, the bottom-right edge has little activity in either cube, and could still go either way. From this point on, many of the units are being strongly enough inhibited by the activations that have been established that they do not get a chance to come on. Only the units in the two bottom-right edges remain in a situation where they would come on if updated. By chance, one of the units in the bottom right edge of cube A is updated before one of the units in the bottom right edge of cube B. This has the effect of making the input to units in the bottom right edge of cube B inhibitory, thereby determining the solution.

Q.3.1.3. Generally, the larger the value of *istr*, the lower the probability of finding one of the two global minima. There are two related reasons. First, units become active more quickly with larger values of *istr*; second, active units exert stronger effects on other units. This means that within a very small number of updates, activations can be set up in each of the two cubes that effectively block completion of the other cube. With smaller values of *istr*, activation is more gradual, and these kinds of blockage effects are less likely.

Q.3.1.4. Providing external input has an effect, but it is not completely overwhelming, for example, providing external input of strength 1.0 biases the outcome, but only slightly. In one set of 40 runs, with *Afll* receiving external input of 1.0, 15 runs settled to cube A and 15 settled to cube B. The bias toward cube A was only apparent in the mixed solutions. There were five runs with six units on from A, four with four units on from both A and B, and only one with six units on from B. Note that the external input is multiplied by the parameter *estr*, which means that external input of 1.0 only contributes 0.4 to the net input. If you turn on two units in the same cube, there is a much more consistent shift toward outcomes favoring that cube. In one set of 40 runs, with *Afll* and *Abur* receiving external input of 1.0, 30 runs settled to cube A, and only 2 settled to cube B.

Ex. 3.2. Schemata for Rooms

Q.3.2.1. Somewhat surprisingly, the system tends to settle to almost exactly the same goodness maxima on different runs with the same unit or units clamped. Note that these maxima are not necessarily global maxima. For example, in the bathroom case, the maximum the network finds has window and drapes off; if they are on (and all else is the same) the goodness is greater. In all but one case, the maxima you should reach correspond to those shown on in Figure 6 in *PDP:14* (pp. 26-27); the only discrepancy is the case of the living room where the unit clamped on initially is *sofa*. The state illustrated in *PDP:14* tends to be reached around cycle 25, with a goodness of about 21.5; but sometime between cycle 25 and cycle 50, the network breaks out of this state and moves toward a higher maximum, with a goodness of very close to 27.0 (it may take a total of 75 cycles or so to converge to this point). This maximum seems to be a luxurious office or a living room with a study area in it.

Differences in the maxima reached are attributable to several factors, the most obvious of which is the number of units that are on. Thus the bathroom prototype, in which only nine units are on, has a goodness of only about 8.08, while the living room prototype, in which 21 units are on, has a goodness of about 27. Another factor, of course, is the pattern of the weights; these determine how much each active unit contributes to the overall goodness. We will leave it to you to discover the effects of clamping various single units and pairs of units.

Q.3.2.2. You will find that the office case gives the clearest example of a situation in which it is better to have both drapes or windows than it is to have either without the other. The bathroom prototype shows this same effect but much more weakly, and in the bedroom and the living room cases, the effect disappears; here having neither is worse than having either, and having both is by far the best. What is happening is that drapes and windows each have net positive input from other features of bedrooms and living rooms, so each contributes positively to the goodness, independently of the other. The cases in which the synergy appears are cases in which each separately gets net negative input from other features that are active, but when the other is active, the balance shifts. You might also notice that drapes without windows generally seems worse than windows without drapes. Can you figure out why? Remember, connections are symmetrical in the schema model!

Ex. 3.5. Simulated Annealing in the Cube Example

Q.3.5.1. You should find a lower density of local maxima—our experience is that they occur only about 10% to 20% of the time. Also, interestingly, only local maxima in which four units are on in each cube seem to "last." That is, the system tends to move away from maxima in which six units are on in one cube and two are on in the other, even at the minimum temperature of 0.5. What would happen if the minimum temperature were reduced—would that have any effect on this tendency?

Q.3.5.2. Initial temperature is pretty much irrelevant in this case unless it is quite low. More important is the gradualness of the annealing. If it is abrupt, there is a very fast transition from approximately random behavior to virtually deterministic behavior; the results in this case (in their most extreme form) approximate the Hopfield net. With slow enough annealing, the network is guaranteed to settle to one of the highest goodness states.

Ex. 3.6. Electricity Problem Solving

Q.3.6.1. Although the network occasionally seems to be stuck in a local maximum at 200 cycles, given the annealing schedule imposed in the *vir.str* file, our experience is that it will almost always eventually find the best answer if run for 300 or even 400 cycles. In this answer the following knowledge atoms are active:

V1 + V2 = VT:	*d-u-s*
R1 + R2 = RT:	*s-u-u*
I * R1 = V1:	*d-s-d*
I * R2 = V2:	*d-u-u*
I * RT = VT:	*d-u-u*

Other knowledge atoms will occasionally flicker on in this state, and occasionally the active ones will flicker off, but most of the time all of the active ones are on, along with one or two others.

Q.3.6.2. Yes, you can replicate Smolensky's findings, but the model is stochastic, so you may have to average over a number of runs to get reliable results. In general it is not possible to go by the results of one or two runs in these stochastic systems to reach your conclusions.

Q.3.6.3. In general, the network tends to settle first on parts of the answer that most directly depend on the units that have been clamped in the problem specification. Other parts of the answer that depend on these results cannot become firmly established until the conclusions that they depend on are themselves firmly established.

CHAPTER 4

Ex. 4.1. Generalization and Similarity With Hebbian Learning

Q.4.1.1. Each row of the weight matrix is a copy of the input pattern, scaled by the learning rate (0.125), and multiplied by the activation of the corresponding output unit, which is 1 or −1. Row 1, then, is just the input pattern times 0.125 since the corresponding output unit has activation 1. Row 2 is just the input pattern times −0.125 since the corresponding output unit has activation −1.

The columns of the weight matrix can be understood in a similar way. That is, each column is a copy of the output pattern, scaled by the learning rate, and multiplied by the activation of the corresponding input unit. Note that these facts follow directly from the Hebb rule, which states that the weight in row i, column j is equal to the activation of input unit j times output unit i times the learning rate.

Q.4.1.2. When a single association has been stored in a pattern associator, the output of the network is always a scalar multiple of the stored output pattern. The value of that scalar is equal to the normalized dot product of the test input with the stored input pattern. Thus, even when the test input seems very different from the stored input pattern, as long as the normalized dot product of the two is positive, the obtained output will simply be a scaled-down version of the stored output pattern. For example, the test input pattern

+ + + + + − + +

has a normalized dot product of 0.25 with the stored input pattern

+ − + − + − + +

The output obtained with this test pattern is 0.25 times the stored output pattern. Thus the *ndp* of the test output with the stored output is 0.25, and the *nvl* of the obtained output pattern is also 0.25. The *vcor* of the test output with the stored output is 1.0, indicating that the vectors are identical up to a scalar multiple.

A special case of this arises when the test input is orthogonal to the stored input pattern. In this case, the normalized dot product of the test input with the stored input is 0, so the test output vector is 0 times the stored output vector; that is, it is a vector of 0s. When the test input is anticorrelated with the stored input, their normalized dot product is negative. When the stored and test inputs are perfectly anticorrelated, their normalized dot product will be -1.0, so the test output will be -1.0 times the stored output. Note in this case that the magnitude of the output (the *nvl*) is just as great as it is when the test input pattern is the same as the stored input pattern.

One useful way of thinking of what is going on in these tests is to note that the output vector can be thought of as a weighted sum of the columns of the weight matrix, where the weights associated with each column are determined by the activations of the corresponding input units. Each of these column vectors is simply equal to 0.125 times the output pattern times the sign the corresponding input unit had in the stored input pattern. When the test input pattern matches the stored pattern, what we end up with is the sum of eight column vectors, each equal to 0.125 times the stored output pattern. The result is 1.0 times the stored output pattern. Different input patterns simply weight the columns differently, but the result is always some scalar multiple of the basic output pattern column vector.

Ex. 4.2. Orthogonality, Linear Separability, and Learning

Q.4.2.1. Your orthogonal set of vectors will be perfectly learned by the net in the first epoch of training, so that during test, the *ndp*, *nvl*, and *vcor* are all 1.0. In vector terms, each test input produces an output pattern that is equal to the sum of the three stored output patterns, each weighted by the normalized dot product of the test input with corresponding stored input patterns. Since the input vectors form a mutually orthogonal set, input pattern k produces an output that is 1.0 times the corresponding output pattern k, plus 0.0 times each of the other output patterns. Additional epochs of training increase the strengths of the connection weights. One can think of this as a matter of increasing the number of copies of each association that are stored in the net.

Each copy contributes to the output. The result is that the *ndp* and *nvl* grow linearly, even though the *vcor* stays the same.

For the set of vectors that are not orthogonal, you should find that learning is not perfect. Instead, the output pattern produced by each test pattern will be "contaminated" by the other output patterns. The degree of contamination will depend on the similarity relations among the input patterns, as given by their normalized dot product, because the output produced by a test pattern is the sum of the outputs produced by each learned input pattern, each weighted by the normalized dot product of the learned input pattern with the test input pattern. Repeated training epochs will not change this. Though the magnitudes of the output vectors will grow with each training epoch, the *tall* command will show the same degree of contamination by other output patterns, as measured by the *vcor* variable.

Q.4.2.2. When the delta rule is used, the results in the first epoch for the orthogonal patterns are the same as with the Hebb rule. Thereafter, however, no further learning occurs because the learning that occurred in the first epoch reduced the error terms to 0 for all output units. For the linearly independent patterns, the benefit of the delta rule becomes clear: the *vcor*, *ndp*, and *nvl* measures will all eventually converge to 1.0. You will note that the last pattern pair in the training list always produces perfect results when you do a *tall*. This is because, given the value of the *lrate* parameter, the error for the current pattern is always completely reduced to 0. Because the changes that produce this effect generally interfere with previous patterns, however, the results are not so good for earlier patterns on the list. If you watch the *vcor* and *ndp* during training, you will note that they stay less than 1.0 for all of the patterns—unless, of course, the pattern set you are using contains a pattern that is orthogonal to both of the others in the set. The values gradually come closer to 1.0 as the changes to the weights get smaller and smaller.

Q.4.2.3. As we saw at the beginning of the chapter, each weight in the Hebb rule is proportional to the correlation between the corresponding input and output units over the set of patterns. You can retrieve these correlations by dividing each weight by the learning rate times the number of pattern pairs stored. In this instance, given that there are three patterns in the training set, each weight has a value of 0.375, 0.125, −0.125, or −0.125, corresponding to correlations of 1.0, 0.33, −0.33 and −1.0. With the delta rule, the weights will tend to preserve the same sign, although they do not have to. The magnitudes will not in general be the same. Basically, the way to think about what is happening is this. The magnitudes of the weights in each row adjust to predict the output unit's activation correctly. If a particular output unit is perfectly correlated with only one of the input units over the ensemble of patterns, the corresponding weight will grow quite large. On the other hand, if a particular output unit is perfectly correlated with a number of the input units, they will come to "share the weight." In the patterns in the *li.pat* file, two units are perfectly correlated with output unit 0 and one is perfectly anticorrelated. These develop weights of +0.29 and −0.29, respectively; these weights actually are smaller than they would be in the Hebbian case. On the other hand, output unit 2 correlates perfectly with only one input unit (input unit 6, the next-to-last one) in these patterns. (Actually, the two units are anticorrelated, but that is equivalent to perfect correlation except a sign change, since one is perfectly predictable from the other.) Thus the weight to unit 2 from unit 6 must be much larger (−0.67) because it cannot share the burden of predicting output unit 2's activation with other input units.

Q.4.2.4. The orthogonal pattern is mastered first because the output it produces is not contaminated by any of the other output patterns.

Q.4.2.5. With Hebbian learning, nothing but correlations, scaled by the number of training trials, get stored in the weights. With the delta rule and *lrate* set at 0.125, the weights oscillate back and forth, falling into a pattern that suffices for each pattern pair in turn, but does not work for some or all of the others. If the learning rate is set to a smaller value, say 0.0125 rather than 0.125, the oscillations are smaller. The weights will average out to values that minimize the total sum of squares.

Ex. 4.3. Learning Central Tendencies

Q.4.3.1. Basically, what happens is that the expected value of each weight converges to the value that it achieved with noiseless patterns in one epoch of training. The higher the learning rate, the sooner the expected value converges, but the more variability there is around this expected value as a result of each new learning trial. Smaller learning rates lead to slower convergence but greater fidelity to the central tendency.

Ex. 4.4. Lawful Behavior

Q.4.4.1. In the particular run that we did, the weights at the end of 10 epochs were:

```
 36-28-16  0 -4 -4 -6 -2
-22 38-24 -2  0 -6 -6 -2
-26-22 38  0  0-10 -6 -4
 -2 -8  0 38-24-24 -6 -4
 -8-12  0-34 38-24 -6-14
 -8 -2 -4-22-26 34 -4-10
  0 -2 -4  0 -4 -2-40 34
 -4  4  0  4 -4  0 38-38
```

Given these values, the pattern

```
1 0 0 1 0 0 1 0
```

would produce a net input of 38 to the last output unit, and

```
1 0 0 1 0 0 0 1
```

would produce a net input of −38. Passing these values through the logistic function with temperature 15 produces a probability of .926 that the last output unit will come on in the first instance and a probability of .074 that it will come on in the second instance.

Q.4.4.2. A third of the time each input unit in each of the first two subgroups is on, a particular unit in each of the other subgroups should go on in the output. The other two-thirds of the time the other output unit should go off. Thus, the weights between input units in one of these subgroups and output units in the other tend to be decremented more often than they are incremented, and that is why the weights from units in the first subgroup (units 1-3) to units in the second subgroup (units 4-6) are negative. Each of the two output units in the 7-8 group, however, is on exactly half of the time each of the input units in the other two groups is on. Thus these weights tend to be incremented as often as they are decremented.

Q.4.4.3. The most probable response to *147* at this point will be *148*, although the tendency to make this response will generally be weaker than the tendency to make response *258* to *257*. The weights from units 1 and 4 to unit 7 should be tending to oppose the strong negative weight from input unit 7 to output unit 7. However, this tendency should not be very strong at this point. The weights that were previously involved in producing *147* from *147* have virtually disappeared. This can be seen by comparing the case in which the weights were not reset before training with the *all.pat* set to another run in which the weights were reset before training with the *all.pat* set. The resulting weight matrices are virtually identical, except for random perturbations due to noise.

Q.4.4.4. It takes so long to get to this point for four reasons. First, the exceptional pattern, *147*, which is the hardest to master, only occurs once per epoch. Second, as the error rate goes down, the number of times the weights get changed also goes down because weight changes only occur when there are errors. Third, several other input patterns (particularly *247*, *347*, *157*, and *167*) tend to work against *147*. Finally, as the weights get bigger, each change in the weights has a decreasing effect on performance because of the nonlinearity of the logistic function.

CHAPTER 5

Ex. 5.1. The XOR Problem

Q.5.1.1. The value of *error* for a hidden unit is the sum of the *delta* terms for each of the units the hidden unit projects to, with each *delta* term weighted by the value of the weight from the hidden unit to the output unit. The *error* for unit 2 (the first hidden unit) is therefore

$$-0.146 \times 0.27 = -0.039.$$

The *delta* for a unit is just its *error* times the derivative of the activation of the unit with respect to its net input, or the activation of the unit times one minus the activation. Thus the *delta* term for unit 2 is

$$-0.039 \times 0.64 \times (1 - 0.64) = -0.009.$$

For unit 3, the *error* is

$$-0.146 \times 0.08 = -0.0117,$$

and the *delta* term is

$$-0.0117 \times 0.40 \times (1 - 0.40) = 0.0028.$$

The values are small in part because the weights from the hidden units to the output units are small. They are also affected by the fact that the *error* is multiplied by the derivative of the activation function with respect to the net input to the unit, first for the output unit, in computing its *delta*, then again for the hidden units, in computing their *delta*s from the *delta*s of the output units.

Each time the error derivative is "passed through" the derivative of the activation function, it is attenuated by at least a factor of 4, since for activations between 0 and 1, the maximum value of *activation* (1 − *activation*) is 0.25. The *delta*s for the hidden units have been subjected to this factor twice.

Q.5.1.2. The network produces an output of 0.60 for input pattern 0 and 0.61 for the rest of the patterns. The results are so similar because the small random starting weights have the effect of making the net input to each hidden unit vary in a rather narrow range (from −0.27 to 0.60 for unit 2 and from −0.44 to −0.36 for unit 3) across the four different patterns. The logistic function squashes these variations considerably since the slope of this function is only 0.25 at its steepest. Thus the activation of unit 2 only ranges from 0.43 to 0.64, while that of unit 3 stays very close to 0.40. This attenuation process repeats itself in the computation of the activation of the output unit from the activations of the hidden units.

Q.5.1.3. The bias term of the output unit and the weights from the hidden units to the output unit all got smaller. There were slight changes to the weights from the input units to the hidden units and in the bias terms for the hidden units but these are less important. Basically, what is happening is that the network has reduced the *tss* somewhat by making each pattern produce an output quite close to 0.50 rather than 0.62. The network achieved this by reducing the weights to the output unit and by reducing its bias term. Note that if the output were exactly 0.50 for each pattern, the *tss* would be 1.0. The *delta*s for the hidden units have gotten even smaller because of the smaller values of the weights from these units to the output unit. Since the gradient depends on these *delta* terms, we can expect learning to proceed very slowly over the next few epochs.

Q.5.1.4. The more responsive hidden unit, unit 2, will continue to change its incoming weights more rapidly than unit 3, in part because its *delta* is quite a bit larger due to the larger weight from this unit to the output unit. Another factor hindering progress in changing the weights associated with unit 3 is the fact that the weight error derivatives for weights both coming into and going out of this unit cancel each other out over the four patterns.

Q.5.1.5. Unit 2 is acting like an OR unit because it has fairly strong positive weights from each input unit, together with a bias near 0. When neither input is on, its activation is close to 0.5. When either unit comes on, its activation approaches 1.0. The other hidden unit remains rather unresponsive, since the weights from the input units to this unit are very small. It is beginning to develop a negative weight to the output unit. Though it is not doing this very strongly, we can see that it is beginning to serve as a unit that inhibits the output unit in proportion to the number of active input units.

Q.5.1.6. Unit 3 is now in a linear range, allowing its activation to reflect the number of active input units. Because it is developing a strong negative weight to the output unit, the effect is to more and more strongly inhibit the output unit as the number of active input units increases. The other hidden unit no longer differentiates between one and two input units on, since its large positive weights from the input units cause its activation to be pushed against 1.0 whenever any of the input units are on. Thus the inhibition of the output unit by unit 3 increases as the number of active input units increases from one to two, and the excitation of the output unit by unit 2 does not increase.

Q.5.1.7. At epoch 240, the *tss* is mostly due to the fact that the network is treating all of the patterns with one or more hidden units on as roughly equivalent. All three patterns are producing activations greater than 0.5. The pattern in which both input units are on (pattern 3) is producing a larger *pss* than the other two of these patterns—it is exerting a powerful effect. In more detail, the weight from unit 3 to the output unit is pushed to be more positive in the two cases in which one input unit is on but is pushed to be more negative when both input units are on. The effects almost cancel but are slightly negative because the activation of unit 3 is larger when both input units are on and the *delta* for the output unit is larger. Meanwhile, unit 3's *delta* is negative in each of the two cases where one input unit is on and is positive when both input units are on. These almost cancel in their effects on the weight error derivatives for the weights from the input units to unit 3, but again the *delta* term for the pattern in which both input units are on dominates. (Remember that each input unit is on in only one of the two patterns in which only one input unit is on.) Thus the net force on the weight from each input unit to unit 3 is positive. As the weight from unit 3 to the output unit gets larger, *delta* for unit 3 gets larger, so the changes in the weights coming into unit 3 get larger and larger.

Q.5.1.8. One could almost write a book about the process of solving XOR, but the following captures most of the main things that happen. Initially, the network reduces the weights from each hidden unit to the output unit and reduces the bias term of the output unit.

This reduces the *tss* to very close to 1.0. At this point, learning slows to a near standstill because the weight error derivatives are quite small. Gradually, however, the hidden unit that is slightly more responsive to the input (unit 2) comes to act as an OR unit. During this phase the total sum of squares is being reduced because the network is moving toward a state in which the output is 0 for the (0 0) input pattern and is 0.67 for the other three patterns. (If it actually reached this point the *tss* would be 0.67.) At this point, the other hidden unit begins to come into play. The forces on this unit cause its input weights to increase and its weight to the output unit to decrease. It comes to inhibit the output unit if both input units are on, thereby overriding the effects of the other hidden unit in just this case.

There is a long phase during learning in which the weight error derivatives are very small, in part because the weights from the hidden units to the output units are small and in part because the hidden units are not very differentiated in their responses and so components of the the weight error derivatives contributed by each of the four patterns nearly cancel each other out. However, between epoch 140 and epoch 220 or so, the gradient seems to be pretty much in the same direction, as indicated by the *gcor* measure. Learning can be speeded considerably if the *lrate* is increased during this period. In fact with the particular set of starting weights in the *xor.str* file, learning can be speeded considerably by setting *lrate* to a much larger value from the beginning. In general, though, high values of *lrate* throughout can lead to local minima.

Ex. 5.4. A Cascaded XOR Network

Q.5.4.1. In cascade mode, the net input to each hidden unit and to the output unit builds up gradually over processing cycles. At first, the net input to all these units is strongly negative because of the negative bias terms that determine the resting levels of these units. When only one input unit it turned on, at first it produces a weaker input to each hidden unit than is produced when both input units are on. This means that unit 2 comes on more quickly when both input units are turned on than when only one is on. Since it is unit 2 that tends to excite the output unit, the activation of the output unit builds more rapidly at first when both input units are on. Note that early on, it makes little difference to unit 3 whether one or both inputs are on, since this unit's bias is so negative. Gradually, however, as processing continues, the activation of unit 2 begins to saturate whether one or both input units are on. During this phase, if both input units are on, unit 3's activation will grow. Because of its inhibitory connection to the output unit, the net input to the output unit will gradually become negative in this case.

Ex. 5.5. The Shift Register

Q.5.5.1. In our experience, the network comes up with two different solutions about equally often. In one of these, the network acts as a normal shift register. Each unit has a strong excitatory connection to the unit to its right; otherwise the connections are usually nearly all inhibitory. In this instance, the pattern called *first output* is shifted one bit to the right from the pattern called *input*, and the pattern called *second output* is shifted an additional bit to the right. In the other solution, the network approximates what might be called a shift-and-invert operation at each step. Thus the first output pattern is shifted one bit right from the input pattern and inverted, so that 1s become 0s and 0s become 1s. This description holds only for the patterns in which one or two of the input units is on, however; for the other two patterns, (no units on or all units on) no inversion is performed. Interestingly, this second solution seems to take longer to reach, even though the two are found about equally often. When the constraint that the biases be negative is removed, the shift-and-invert solution is purer and easier to achieve, apparently at the same time making the network slower to converge on the simple shift solution. If the links between the weights are removed, there are a very large number of different solutions. Interestingly, it seems to take longer in general to solve the problem in this case. This is one of many examples in which constraining a **bp** network improves its ability to solve a particular problem.

Ex. 5.6. Plan-Dependent Sequence Generation

Q.5.6.1. The problem is very hard because the input patterns that the net must learn to respond differentially to do not start out very different. As learning proceeds, they gradually differentiate. As this occurs, the network must adjust itself to these changes. You can compare the "copy-back" version of this problem to the case in which the current input is supplied as part of the input by editing the *seq.pat* file so that the negative integers that cause the copy back are replaced by the values of the target from the previous line. In this case the problem is usually solved more quickly.

CHAPTER 6

Ex. 6.1. The Linear Hebbian Associator

Q.6.1.1. The receiving unit for the weight in row 3, column 0 is unit 3; the sending unit is unit 0. The values in row 3 reflect the product of the activation of the receiving unit, the activations of each of the sending units, and the *lrate*, which is 0.125. Since the activation of the receiving unit (unit 3) is −1.0, row 3 is −0.125 times the activation vector.

Q.6.1.2. The two learning patterns *a* and *b* are orthogonal to each other. The internal input to a receiving unit is equal to the dot product of the current activation vector with the vector of weights on the connections to that receiving unit. This vector is equal to *lrate* times pattern *a* times +1 or −1, depending on the particular receiving unit in pattern *a*. Regardless of the sign, the dot product of the current activation vector with the vector of weights is 0 because the two patterns are orthogonal.

Q.6.1.3. The weights in row 2 are 0.125 times pattern *a* plus −0.125 times pattern *b*. Patterns *a* and *b* differ in elements 1, 2, 4, and 7; *a* and −*b* differ in elements 0, 3, 5, and 6. Row 2 has 0s in every element where the two base patterns agree, or where $a_i - b_i$ is 0.0. In other cells of this row, the values do not cancel out; they are either +0.25 or −0.25.

Q.6.1.4. On cycle 1, the external input is applied, so that the activations of the units at the end of this cycle reflect the external input alone. On cycle 2, these activations, together with the weights, produce an internal input that exactly reproduces the external input pattern. These two patterns together produce the resulting activation, which is 2 times the external input in each case. These results follow from the fact that the response to each external input vector (that is, the internal input generated by it) is equal to *k* (or *nunits* times *lrate*) times the sum of the stored vectors, each weighted by the dot product of itself with the external input.

Q.6.1.5. Before training with the set of four patterns, the two that have previously been learned will produce an internal input that matches the external input, and the two new ones will produce no internal input because they are orthogonal to the stored patterns. After one epoch of training, the two patterns that had previously been presented produce internal input equal to twice the external input, and the patterns that have only been presented once produce internal input exactly equal to the external input. Essentially, each pattern in the set is an eigenvector of the network. The eigenvalue is equal to the number of presentations times k (or *nunits* times *lrate*). Regarding a negated vector that is -1 times a stored vector, any scalar times an eigenvector of a matrix is still an eigenvector of the matrix, so the magnitude of the response to the negated vector is equal to the magnitude of the response to the vector from which it was derived.

Q.6.1.6. For patterns that have been learned only once, the internal input at time cycle t is equal to the prior activation. When this is added to the continuing external input, the resulting new activation increases by *estr* times the external input. For patterns that have been learned twice, the internal input is twice the previous activation value so activations grow much more quickly. This explosive growth would continue indefinitely except that there is a built in check that stops processing as soon as the absolute value of the activation of any unit exceeds 10.0.

Q.6.1.7. The weight matrix that results from learning these patterns will be the identity matrix. These weights will produce an internal input that matches the current pattern of activation for any pattern of activation. In terms of the weights, this is because the weight from each unit to itself is 1 and the weight from each unit to every other unit is 0. In terms of the eigenvectors, any vector in n-dimensional space can be written as the weighted sum of any set of n orthogonal vectors in that space; such a set of vectors *spans* the space. Since the eigenvalue of each of these vectors is 1.0 for the matrix you have created in this exercise, the vector that comes out is the same weighted sum of these eigenvectors as the vector that went in.

Ex. 6.2. The Brain State in the Box

Q.6.2.1. Each of the trained patterns is an eigenvector of the weight matrix with eigenvalue 2.0, because each pattern was presented to the network twice during learning. At cycle 1, the external input sets the activation of each unit to *estr*, or 0.1 times the external input to each unit. Since the resulting pattern of activation is just 0.1 times the pattern presented, this pattern of activation is an eigenvector, and so, on cycle 2, the internal input pattern will be equal to 2 times the prior activation, or 0.2 times the external input. This product, plus 0.1 times the external input, results in an activation of 0.3 times the external input. On cycle 3, this new activation pattern is again multiplied by 2 by the matrix, so that now the internal input is 0.6 times the external input; when 0.1 times the external input is added to this, the activation becomes 0.7 times the external input. On the next cycle, the doubling occurs again, but this time, 0.1 times the external input plus the internal input is 1.4 times the external input. This would mean that the activations of the individual elements would go outside the "box" imposed by the BSB model. The model does not allow this to happen; it "clips" the activations at +1.0 or −1.0. The next cycle generates an even larger internal input (2.0 times the external input pattern), but this is again clipped; from this point on, the internal input and activation do not change; the pattern of activation has reached a corner of the box.

Q.6.2.2. The pattern +−+−. . . . has an *ndp* of 0.5 with pattern *a* and an *ndp* of 0.0 with pattern *b*. The pattern of activation this produces on cycle 1 has *ndp* of 0.05 with pattern *a*. Thus, the network responds to this pattern by producing as its output (that is, by setting the internal input to the units) to $(2.0)(0.05)a = 0.1a$. The factor of 2.0 is the eigenvalue of *a*. This internal input, when added to 0.1 times the external input, produces a pattern of activation on cycle 2 that has an *ndp* of 0.15 with pattern *a* and still has an *ndp* of 0.0 with *b*. This in turn produces an internal input pattern equal to $0.3a$. Again the external input is added, but already it is quite clear that the pattern of activation is moving in the direction of *a*. This process continues until the activations of the units reach the boundaries of the box. At this point they are clipped off, and all of the units have activations of equal magnitude and replicate pattern *a* exactly.

Q.6.2.3. At the end of cycle 1 the activations are simply 0.1 times the external input. This pattern has *ndp* of 0.075 with *a* and of 0.025 with *b*. This generates an internal input of $0.15a + 0.05b$. The patterns agree on units 0, 3, 5, and 6 and disagree on the other units. Internal inputs to units where they agree have magnitude 0.20 (0.15+ 0.05), whereas internal inputs to units where they disagree have magnitude 0.10 (0.15− 0.05). As it happens, the internal input to unit 7 is −0.10, which cancels the external input (when scaled by the *estr* parameter) so that the activation of unit 7 is 0 on this cycle. The activation of unit 6, on the other hand, is −0.3, because the external input and both learned patterns agree on the activation of this unit. The resulting pattern of activation on cycle 2 has an *ndp* of 0.225 with *a* but only 0.075 with *b*. In general, before the clipping threshold is reached, the pattern of activation will have an *ndp* of $(t-1)(0.15) + 0.075$ with *a* but only $(0.05)(t-1) + 0.025$ with *b*; Thus, *a* will always be three times stronger than *b* in the internal input, with the internal input to unit 6 having twice the magnitude of the internal input to unit 7. The internal input to 6 will always have the same sign as the external input to this unit, but the internal input to 7 will have the opposite sign compared to its external input. On cycle 3 the internal input becomes stronger than *estr* times the external input to 7, so the activation of unit 7 changes in sign at this point. After one more cycle, the activation of unit 6 reaches the clipping threshold, while the activation of unit 7 still lags. After an additional cycle, even unit 7 reaches the clipping threshold. At this point, all of the units have activations equal to their value in pattern *a*.

Q.6.2.4. The network "rectifies" the distorted unit from the prototype after a single pass through the weights, so that by cycle 2, the sign of every unit in the activation vector matches the prototype value. However, the magnitude is less than it would be had the prototype actually been presented, since the distorted unit opposes the other units in the computation of the internal input to each unit. A further factor is that the external input to the distorted unit opposes its rectification directly. The activations of the nondistorted units reflect the first of these two factors, but not the second, so they show a less marked decrement compared to the prototype case. In any case, both inputs end up in the same corner of the box; the network categorizes each distortion by distorting its representation toward the prototype. With regard to the weights, over the ensemble of distortions, each unit correlates perfectly with itself and has a correlation of +0.5 or −0.5 with each other unit. These correlations are reflected directly in the values of the weights.

Ex. 6.3. The Delta Rule in a Linear Associator

Q.6.3.1. Essentially, what usually happens is that the delta rule finds, for each unit, the set of other units whose external inputs are perfectly correlated (positively or negatively) with its own. These correlated units eventually get all of the weight; the units whose activations are not perfectly correlated have no influence on each other. This would not happen if the external input to one of the units was not perfectly correlated with any other unit. When there are perfectly correlated units, the weights coming into each unit have absolute values that sum to 1.0 asymptotically because the delta rule is trying to set the internal input to each unit to match the external input, which is +1.0 or −1.0. This means that in cases where a unit's external input is perfectly correlated with the activation of more than one other unit, the correlated units divide the weight equally among themselves. The weight matrix obtained with the Hebb rule is similar, but as usual with Hebbian learning, the weights reflect pairwise correlations only. Thus, the weight does not gravitate toward the inputs that predict each unit perfectly nor is there any restriction that the weights coming into a particular unit sum to 1.0.

Q.6.3.2. The file *imposs.pat* in the **aa** directory contains a set that cannot be learned in an auto-associator. We reproduce these patterns here:

```
xua  +--++--+
xvb  +--++-+-
yub  +-+-+---
yva  +-+-+-++
```

These patterns were set up so that the last unit would not be predictable from any of the others. The first four units were set up as one subset with two alternative patterns (+––+) and (+–+–) called *x* and *y*, respectively. The next three units were set up as a second subset with two alternative patterns (+––) and (+––) called *u* and *v*. The final unit was set up as a single-unit subpattern with two alternatives (+) and (–) called *a* and *b*. The value of the final subpattern cannot be predicted from either of the other subpatterns, because *x* and *y* each occur with *a* and *b*, as do *u* and *v*. It is only the particular conjunction of the first two subpatterns that can predict the correct activation for the final unit, but the delta rule must be able to set the internal input to each unit based on a weighted sum of its inputs.

Q.6.3.3. Two units that predict each other perfectly will develop strong weights, and if they cannot be predicted by the others, their incoming weights from all other units will be 0. This means that although the auto-associator may be able to learn a set of training patterns perfectly, it may not be able to perform pattern completion accurately if the portion of a known pattern that is used as a probe contains any member of a mutually predictive (and not otherwise predictable) pool. This tends to limit the circumstances in which pattern completion will be possible. In fact, if two units perfectly predict each other, the delta rule will pile all the weight coming into each on the connection from the other, even if the activations of these units could be partially predicted by a weighted sum of activations of other units. A similar problem can arise even in auto-associators that use hidden units. To avoid this problem, it is necessary to introduce noise that keeps activations from being perfectly correlated.

Ex. 6.4. Memory for General and Specific Information

Q.6.4.1. Your weights should look about the same as those in the figure, although, of course, they will differ in detail because you will have used a different random sequence of distortions. You should find that the model responds best to the prototype, and that, on the average, it responds better to the exemplar that it just saw than to new random distortions. Thus, over the short term, it is sensitive to what it has recently seen, but over the longer term, it extracts the prototype and comes to respond best to it.

Q.6.4.2. When you increase *pflip*, you increase the variance of the individual training exemplars from the prototype and from each other, and this increases the extent to which the model acts as though it is retaining information about particular exemplars. The discussion of Whittlesea's (1983) experiments in *PDP:17* (pp. 199-205) draws on this aspect of the model's performance. Increasing the *lrate* parameter also increases sensitivity to recent exemplars at the (slight) expense of the prototype. This is used to account for some aspects of spared learning in amnesia in *PDP:25*.

Q.6.4.3. You should find that the individual weights in your simulation tend to be larger than in the figure. The reason is primarily that the network must rely on the connections internal to the visual pattern, whereas the weights shown in the figure were obtained when it was able to rely on the name pattern as well as the visual pattern.

Q.6.4.4. You should find more adequate completion with *ncycles* equal to 50. If you round off the obtained activations (so that 0.16 rounds up to 0.2, for example), you should get results fairly similar to those shown in Table 1. The model seems to do a good job filling in what it can but does have one limitation that makes it different from human performance: In cases where it is probed with an ambiguous probe, it always fills in a blend of the two responses that would be appropriate rather than selecting one or the other response.

Q.6.4.5. At the end of 50 epochs, it appears that learning has not yet quite finished, and so some of the elements of the retrieved patterns are occasionally incorrect. In our runs we get the correct results after 100 training epochs.

The network has the most trouble with those properties that differ between typical instances and specific familiar examples. We might expect human subjects to have similar problems, confusing the properties of typical and familiar exemplars. However, once again it seems likely that humans would eventually learn to differentiate the specific instance from the prototype somewhat more fully than we seem to observe here. It is likely that stochastic auto-associators with hidden units would be necessary to capture these characteristics adequately in a PDP framework.

Ex. 6.5. Clustering the Jets and Sharks

Q.6.5.1. The typical pattern of weights indicates that each Jet is activating one of the two output units and each Shark is activating the other. The system usually falls into this pattern because it minimizes the differences between the patterns within each cluster. On average, the Jets are more similar to each other than they are to the Sharks. This similarity is in part because the Jets all have the Jets feature and the Sharks all have the Sharks feature.

Q.6.5.2. The pattern of weights over the other features represents the distribution of values for the Jets and the Sharks. Each dimension gets about 20% of the weight and divides this up so that each value on the dimension has a weight whose average is proportional to the frequency that this value occurs. The observed weights will fluctuate a bit since they are adjusted as each pattern is presented. In the particular case of the *Age* units, most Jets are in their 20s and most Sharks are in their 30s, so the 20s connection gets most of the weight to the unit that responds to the Jets and the 30s connection gets most of the weight to the unit that responds to the Sharks.

Q.6.5.3. Many different possible nonoptimal configurations of weights can be observed. Generally, each one involves one or perhaps two dimensions. For example, the following weights were produced on one run:

```
10  9          9 10
 5 11  3       9  8  2
 5  8  6       9  8  2
12  6  1       4  9  5
10  0  9       0 19  0
```

Here the network has partitioned the patterns in terms of whether they are Burglars (second feature on last row) or not. From this we can see that the non-Burglars (who activate the left unit) tend to be in their 30s and single, whereas the Burglars tend to be in their 20s or 30s, with LH or HS educations and with about half of them married. Note that this is a much less optimal partitioning than the typical one.

Reducing the learning rate makes it easier to find an unusual set of weights because the learning process is doing gradient descent in the compactness of the partitions, but each step is being made on the basis of a very small sample (a single pattern). Learning can be thought of as tending downhill, but with noise so that it does not always go downhill. When the *lrate* is higher, that is like more noise, and the network can escape from local minima with greater probability. Note that as the weight step gets infinitesimally small, each weight step will have very little effect and so performance will approximate true gradient descent more and more closely. The nonoptimal states are local minima due to the initial configuration of the weights. As the weights are reorganized, it becomes harder and harder to break out of the current minimum; that is why the pattern of weights eventually gets locked in.

Q.6.5.4. The network no longer categorizes the groups into Jets and Sharks because the Jets and Sharks overlap quite a bit and there are several better ways they might be partitioned into groups. One example case is shown below:

```
 U  U          U  U
21  0  3       0 21  3
12  8  4       6 11  6
14  6  4       9 12  3
 7 10  6       7  8  8
```

Here the network has used age (20s vs. 30s) as the categorization feature, dividing up the 40s individuals in terms of their similarity to the members of the other two age groups. This partition is quite like the Jets vs. Sharks partition, but it groups Ken, a Shark with two of the three typical Jet properties (as many as any Jet has), with most of the Jets, and groups Mike, Al, Doug, and Ralph, the 30s Jets, with most of the Sharks.

Q.6.5.5. When the gang categorization feature is present (Jets vs. Sharks) the network tends to construct one cluster consisting of all the members of one gang, and then tends to divide the other gang into two subgangs. Because there are more Jets than Sharks, it is more typical for the Jets to be divided into two groups, since this gives a slightly more even distribution of individuals in each of the three groups, but this does not always happen. In fact, the network will occasionally find a group of similar individuals who are about half Jets and half Sharks, putting the rest of the Sharks in one group and the rest of the Jets in the other. There is one sensible pattern that is quite common when there are three output units and the *ajets.pat* patterns are used so that the gang feature is not specified. This is the pattern in which each unit picks up on the individuals with a different occupation:

```
 0  0           0  0           0  0
12  9  2        8 14  1        5 11  7
12 10  2        4 11  8       10  9  5
 4 12  7       13  8  2       19  5  0
 0 24  0       24  0  0        1  0 23
```

This division of the individuals turns out to reveal that there is some systematicity to the distribution of other features, dependent on the occupation feature. For example, the algorithm discovers that the bookies tend to be the least educated.

Ex. 6.6. Graph Partitioning

Q.6.6.1. In 9 out of 10 runs that we tried with the default value of *lrate*, the best solution was found; that is, one unit responded to all patterns involving two units in the left part of the graph, and the other responded to all patterns involving two units in the right part of the graph. The single pattern that connects the two parts of the graph is grabbed by one of the two units. On the one exceptional run, the network found the following local minimum:

```
         7          13              5          12

18         0  0        32      0       21  37        0

        10          18              6          16
```

In this state, the two inputs that have the largest number of connections (those that connect the two graphs) form a strong "bond"—that is, they both activate the same output unit. This leaves the other output unit to pick up the patterns that do not activate either of the linking input units. When smaller *lrate* is used, this and several other local minima can be found more frequently, for the same reasons as discussed in Q.6.5.3.

CHAPTER 7

Ex. 7.1. Using Context to Identify an Ambiguous Character

Q.7.1.1. The letter unit *k* begins to get an edge over *r* at cycle 3, one cycle after *work* exceeds threshold, but the effect at this point is in the third decimal place, since the activation of *work* is very slight (less than 0.01) at the end of cycle 2. A detectable advantage for *k* appears on the screen at the end of cycle 4. The advantage is due to feedback from *work*, and it grows larger and larger as *work* becomes more and more strongly excited. The parameter that determines whether the activation of *r* will decrease as the activation of *k* increases is the letter-to-letter inhibition parameter *gamma/ l->l*. The response probability for *r* eventually goes down because the response-choice probability for a particular alternative is based on the Luce (1959) ratio, in which each alternative's choice probability is equal to its own response strength divided by the sum of the response strengths of all of the other alternatives. As the response strength for *k* goes up, its contribution to the denominator increases, and therefore the choice probability for *r* goes down. The response probabilities pull apart more slowly because response strengths build up gradually over time, based on the *orate* parameter. If this parameter is set to a larger value than its default value of 0.05, response probabilities shadow activation differences more closely. The asymptotic value is unaffected, however (although with a low *orate* it takes quite a while to reach asymptote).

Ex. 7.2. Simulations of Basic Forced-Choice Results

Q.7.2.1. For the *CAVE* display, the activation of *v* starts at 0 and grows by about 0.06 or 0.07 on each cycle, slowing down at around cycle 8 and reaching a level of 0.73 on cycle 15. When the mask comes on in cycle 16, its effect is sudden, wiping out the activation of *v* in two cycles. The forced-choice results are based on the response probabilities. The response probabilities for *v* and *g* start out equal at .03 (actually .038, or 1/26), since all letters start out with an equal chance of being chosen before the onset of the display. The advantage for *v* begins to appear immediately, but builds up more slowly at first because of the low value of *orate*. By cycle 12 it is growing quite rapidly, however, and reaches a peak of 0.60 on cycle 15. It falls off much more gradually than the activation of *v*, again because of the low value of *orate*. The forced-choice accuracy at the end of processing is based on the response probabilities at the point in processing where the response probability for the correct alternative is maximal, which is at cycle 15, just before the onset of the mask. The probability of choosing *v* in the forced choice is equal to its response probability at cycle 15 (.60), plus 0.5 times the probability that neither *v* nor *g* is read out (about .38), which comes to .60+.19=.79.

For the *ZXVJ* display, the time course of activation of *v* is very similar at first, but *v* rises more slowly after cycle 3, reaching only 0.44 at cycle 15. The response strength for *v* is correspondingly lower, and the resulting forced-choice accuracy reflects this.

The *ZXVJ* display produces no word-level activations because of letter-to-word inhibition. The letter-to-word inhibition is 0.04 while letter-to-word excitation is 0.07. Given these values, if *ZXVJ* had two letters in common with some word, that word would receive a net excitatory input because two letters in the display would activate the word by 0.07 times their activation and two letters would inhibit it by 0.04 times their activation, for a net input of about 0.06 times the letter-level activation. Since *ZXVJ* has at most one letter in common with any word, the set of active letters produces at best a net inhibitory effect of $0.07-(3\times 0.04)=-0.05$ times the letters' activations.

No words other than *cave* achieve substantial activation when *CAVE* is presented because of the strong word-to-word inhibition. The word-to-word inhibition parameter is set at 0.21. At first all words with two or more letters in common with *CAVE* receive net excitatory input from the letter level, but as the activations of these words exceed threshold they inhibit each other. The word unit *cave*, which receives by far the most bottom-up excitation, continues to grow in activation, but the others grow very slowly or not at all. As *cave* gets more active it eventually comes to suppress the activation of the others; they cease inhibiting *cave*, so its activation grows with little restraint, thereby allowing it to increase its domination of the pattern of activation at the word level.

Q.7.2.2. Accuracy for *V* in *MAVE* is almost as good as for *V* in *CAVE* because *MAVE* activates a number of words containing *V* in the third letter position. Though none of these words achieve the level of activation *cave* achieves when that word is presented, all of the active words containing *V* work together, providing feedback that reinforces the activation of *v* in *MAVE*.

Q.7.2.3. The reason the advantage of *have* grows larger and larger is that it exerts more of an inhibitory influence on its neighbors than they exert on it. The *have* word unit's initial resting-level advantage means that it crosses threshold sooner and has a larger output than other words receiving an equal amount of excitation from the letter level. This means that other words see more inhibition from *have* than *have* sees from them. In McClelland and Rumelhart (1981) this is called the "rich get richer effect." (This effect is similar but not identical to Grossberg's rich get richer effect described in Chapter 2.)

What happens with *move* is even more interesting. Here, though *move* is one of the most frequent words, the letters that it has in common with the display are shared with fewer of the other words that have three letters in common with the display. The words *cave*, *gave*, *have*, *pave*, *save*, and *wave* all have *_ave* in common with the display, while *made*, *make*, *male*, *mare*, *mate*, and *maze* all have *ma_e* in common with the display. The word *move* is the only word that has *m_ve* in common with the display, and no words have *mav_* in common with the display. Thus, 7 words (*move* and the *ma_e* gang) are providing feedback support to *m*, 12 words (the *_ave* gang and the *ma_e* gang) are supporting *a*, 7 are supporting *v* (*move* and the *_ave* gang), and all 13 are supporting the final *e*. The result is that *a* and *e* are more strongly activated at the letter level than *m* or *v*. Thus the members of the *_ave* gang and the *ma_e* gang get more bottom-up activation than *move* as a result of the differential top-down feedback. Word-to-word competition causes *move*'s activation to fall off as these other words gain.

Q.7.2.4 Examples of these effects can be obtained with almost any nonword, but they work better for nonwords that have three letters in common with several words. A nice example of the gang effect can be seen with *RARD*, in which *rare* loses out to the *_ard* gang.

Ex. 7.3. Subtler Aspects of the Word Superiority Effect

Q.7.3.1. High bigram-frequency pseudowords like *TEEL* tend to activate more words than low bigram-frequency pseudowords like *HOET*, but for these examples, both kinds tend to activate only words that have the correct forced-choice alternative; they differ from the word displayed in some other position. Thus at cycle 15, the words activated by *TEEL* are *feel, heel, keel, peel, reel,* and *tell* (*teen* has a tiny residual activation). The summed activation of these words is the feedback support for the *L* in *TEEL*. The sum is about 0.64 (0.62 shows in the display due to truncation). With *HOET*, though fewer words have three letters in common with it, all end in *T*, and at the end of 15 cycles, they are the only words left active, with a summed activation of about 0.66. Thus the lower bigram-frequency item has, in fact, a bit more support than the higher bigram-frequency item.

Q.7.3.2. McClelland and Johnston got the results that they did (according to the interactive activation model) because their method of constructing pseudowords caused the target letter to benefit from the support of words partially activated by the display. As it happened, both for high and low bigram-frequency words, the number of "friends" of the target letters (i.e., words that had three letters in common with the display, including the target letter) was very high and there were virtually no enemies of any of their low bigram-frequency words. (An "enemy" is a word that has three letters in common with the display, in which the three letters do not include the target letter tested in the forced choice). When an item like *HOEM* or *HOET* is tested in the position of the changed letter, the bias in favor of friends is reversed. Low bigram-frequency items are particularly susceptible to these effects, and so if McClelland and Johnston had tested the changed letter they might have observed a large apparent bigram-frequency effect.

Q.7.3.3. In the bright-target/pattern-mask condition, the more highly constrained word usually has an advantage at the word level due to reduced competition. The correct answer always dominates, however, and the word-level difference makes a smaller difference at the letter level for two reasons. For one thing, the forced choice tends to attenuate word-level differences. More interestingly, words with weaker constraints tend to have more friends as well as enemies. That is, while there are more words that share three letters with the display and differ from it in the tested position, there are also more words that share three letters with the display and match the display in the tested position (e.g., *CAME*, *CARE*, and *CASE* in the case of *CAKE*). These words support the target letter, thereby counteracting some of the disadvantage these words would otherwise have because of reduced feedback support from the target word itself.

Under degraded display conditions, the features extracted are much more likely to be compatible with words other than the word shown if the word shown is a low-constraint word than if it is a high-constraint word. For example, with *CAKE*, the words *LAKE* and *CARE* are frequently consistent with the features detected; *CAFE* and *CAPE* occasionally are as well. With *CLUE*, on the other hand, the only other word that we have ever found consistent with the display is *GLUE*, and this occurred only once in about 20 runs. Note that, when some other word is consistent with the display, the alternatives that are consistent generally do not have an equal response probability; rather, the more frequent alternative tends to dominate. Thus *CAKE* tends to lose out to *LAKE* and *CARE*, both of which are higher in frequency than *CAKE*. These effects tend to greatly increase the advantage of more constrained words under these visual conditions compared to bright-target/pattern-mask conditions.

Ex. 7.4. Further Experiments With the Interactive Activation Model

Q.7.4.1. Both *SLNT* and *SLET* produce a facilitation for *l* relative to *JLQX* because both activate words that share three letters with them, including the letter *l*. *SLNT* also activates *sent*, which does not support the *l*; this is why there is slightly more facilitation for the *l* in *SLET* than for the *l* in *SLNT*. Over the set of materials used by Rumelhart and McClelland, there was a tendency for the pronounceable items to show an advantage in the second and third positions, both in the experimental data and in the simulations. In the simulations, the effect was due to vowel enemies of the consonant target letter.

What is critical is that the target letter have more friends than enemies. For example, if we test *SLNT* in the third letter position, we get very little facilitation because the *N* has only one friend. The *E* in *SLET* does even worse since it has no friends among the words of four letters.

Q.7.4.2. The model produces a very large contextual enhancement effect for words, but the effect is negligible for pseudowords. What happens with pseudowords is that words that match two or more letters from the preview of the context become activated by the context, and they tend to interfere with the later activation of words that specifically share the target letter with the context. This effect can be particularly severe if (as is not the case with *_hig*) all three letters of the context happen to match some word; in this case the model produces a negative pseudoword context enhancement effect; that is, the model does worse on such items when the context precedes the pseudoword than when it does not.

Q.7.4.3. To produce the context enhancement effect with pseudowords, the main thing we did was lower the resting activation level for words (the minimum activation for words had to be reduced to accommodate this). Lowering the resting level for word units had the effect of allowing the preview to prime all words consistent with it, without allowing any of them to go above threshold during the preview period. This meant that when the target letter was presented, the words that included this letter would tend to be the first to exceed threshold—this would then allow the target letter to suppress its competition. By itself this manipulation was not quite sufficient; we found we also had to set the threshold for word-to-word inhibition above 0 so that active words could feed activation back to the letter level before they began to inhibit each other. Finally, we also found we needed to reduce the letter-to-word level inhibition. With these changes, we were able to capture the pseudoword enhancement effect as well as all the rest of the findings considered in McClelland and Rumelhart (1981) and Rumelhart and McClelland (1982). These parameters produce a rather exaggerated enhancement effect with this particular item, but over the ensemble of materials used in the pseudoword enhancement effect experiments reported in Rumelhart and McClelland, they produced only a slightly oversized enhancement effect.

Q.7.4.4. To produce the effects that Carr et al. observed, one has to assume that subjects have control over the letter-to-word inhibition parameter *gamma/ l−>w*. If letter-to-word inhibition is strong, then words other than the word actually shown are kept from receiving bottom-up excitation. This would be appropriate if subjects were expecting only words, since it would tend to reduce interference at the word level. This might also be appropriate when expecting only random letter strings, since it would tend to reduce spurious feedback from partially activated words; alternatively, the letter-to-word inhibition might be set high by default. In processing pseudowords, however, a lower setting of letter-to-word inhibition would allow these stimuli to produce partial activations of words, thereby producing a facilitation effect for letters in these stimuli. If the letter-to-word inhibition is assumed to have a value of 0.21 when words or random strings are expected, but is assumed to be reduced to 0.04 only when pseudowords are expected, the model nicely simulates the Carr et al. effects.

APPENDIX F

An Overview of the PDP Software

In this appendix we provide a brief description of the PDP software, focusing particularly on the user interface and the data structures used by the programs. This description stands in lieu of detailed comments in the source code itself. We assume familiarity with the C programming language, as described in Kernighan and Ritchie (1978). Descriptions of the core routines that define each model are given in the appropriate chapter of the book.

Command and Variable Tables

The user interface is built around a table called the *command table*, in which each command string that the user might enter is stored. The command table contains the name of the command, the menu it is in, a pointer to a function to run if the command is entered, and a pointer to an argument to give to the command. The function *do_command* takes the command entered on the keyboard, searches through the table until it finds a match that is in the current menu, then executes the function associated with the command. If the command is the name of a menu of subcommands, then the function that is executed is *do_command* again, with its argument being the menu associated with the command. When the program is initialized, a number of commands are installed in the command tables, generally in functions whose names begin with *init_*.

There is also a variable table. Each entry in the variable table begins with a string that is used as the name for the variable. This string is used for accessing the variable from the user interface and does not need to be the same as the symbolic name used by the programmer in the program itself. Each entry also contains a variable type identifier and a pointer to the variable. The types include scalar variables (characters, integers, and floating-point numbers), strings (sequences of characters), vector variables (vectors of integers, floats, and strings) and matrix variables (here treated as lists of vectors of either integers or floats). For vector variables, there is an entry specifying the length of the vector. For matrix variables, there is one entry specifying the length of the list of vectors and another entry specifying the length of each vector. There is also a special type of variable called a weight-matrix variable, which will be described specifically later on.

Part of the initialization process involves installing variables as well as commands. Actually, when a variable is installed in the variable table, it is also installed in the command table, generally in the *set/* menu or a submenu of that menu. For example, the variable *nupdates* in the **cs** program is installed as a command in the *set/* menu. This command, when executed, calls the function *change_variable* with an argument that is a pointer to the table entry for *nupdates* in the variable table. The *change_variable* function, in turn, calls a function appropriate for the particular type of variable, as determined from the value of the type field of the entry for *nupdates*.

The Template List

In addition to the command and variable tables, there is also a template list. The template list has in it an entry for each template encountered in the *.tem* file; these entries are installed as the template file is read. Each template list entry has a string that is the external name by which the template can be assessed by the user, a specifier indicating what type of template this is (e.g., vector, matrix, label_array, etc.), a pointer to the variable associated with the template (this variable must be installed in the variable table before the template can be used), and a large number of other variables specifying the parameters of the template, including its screen location, display level, and so on. Once the templates are installed, they govern the process of displaying information in the display area, either when the *update_display* routine is called or when the user issues a command to display a particular template, as when using the *disp/* command; each template name is in fact installed in the command table in the *disp/* menu, with a pointer to the associated entry in the template list as the argument. Each template is also installed in the *disp/ opt/* menu, so that some of the parameters associated with the template can be modified by the user during running.

Matrices as Lists of Pointers to Vectors

Most of the variable types used in the programs are very standard, but two are somewhat special in the way we have implemented them. Matrices are not implemented in the standard way as two-dimensional arrays, but as lists (or vectors) of pointers to vectors. The entry *a[i][j]* refers to the *j*th entry in the *i*th vector, and, in cases where the vectors are all the same length, this notation is functionally equivalent to a conventional two-dimensional array. Weights, however, are stored in lists of pointers to vectors, in which the length of the vectors need not always be the same.

Indeed, the weight arrays are built around the idea that the vector of weights associated with each receiving unit need only be as long as the number of units that project to it.[1] To implement this, there are special arrays called *first_weight_to* and *num_weights_to*. These list the first unit that projects to each unit and the number of units that project to it. To the user, the connection to unit *i* from unit *j* is thought of as *weight[i][j]*, but internally it is stored in the element *j* − *first_weight_to[i]* of the vector *weight[i]*, and so its true index is *weight[i][j − first_weight_to[i]]*. We stress that from the user's point of view, a particular weight is indexed just as it would be in a full two-dimensional array.

The *first_weight_to* and *num_weights_to* arrays are determined by the % specifications found in the network specification files. When no % specification is given, the program sets *first_weight_to* to 0 for each unit and sets *num_weights_to* to *nunits*; thus the weight array in this case is functionally equivalent to a two-dimensional array. The only proviso is that the vectors of weights are not necessarily contiguous.

Extensive Use of Malloc

All of the programs can be configured for networks of arbitrary size and connectivity, limited only by the total available memory in the user's computer. Likewise, the number of variables, patterns, templates, and commands are arbitrary. This is achieved by using the function *malloc* to allocate memory for the various tables, vectors, and lists of pointers to vectors. Initially, many of these data structures are given an arbitrary, reasonable length (e.g., the command table is initially set up for 100 commands). If more entries are needed, a larger number of entries is allocated, the existing entries are copied into it, and the old list of entries is freed for later reallocation. Many of the data structures used by the program cannot be

---

[1] Actually, *num_weights_to* is equal to the number of units from the first unit that projects to the receiver to the last one that projects to it; there could be null weights between the first and the last, but such null weights have entries in the weight vector.

allocated until the network configuration has been defined. For this reason, the programs all have a routine called *define_system*, which is called when the user enters a command that initiates processing. Once called, a variable called *System_Defined* is set to 1, indicating that *define_system* does not need to be called again.

An Overview of the Structure of the Programs

Each program consists of a set of basic parts, each found in a separate file. The core of each program is located in a file whose name is the name of the program followed by *.c*; thus the guts of **bp** is in *bp.c*. These core files contain the routines that define the actual computations performed by the model and that compute statistics relevant to the model (e.g., *goodness*, *harmony*, or *tss*). The other parts are shared by most or all the programs. One part is concerned with the installation of commands and the routines that read commands and execute the appropriate functions as a result; it is located in the file *command.c*. Another part is concerned with the installation of templates; it is located in the file *template.c*. The file *display.c* contains the routines used to display the templates. The file *variable.c* contains functions relevant to the installation and modification of the entries in the variable list. The *weights.c* file contains routines for reading network specification files and for reading and writing weights to files. The file *patterns.c* contains routines for reading patterns and pairs of patterns into the program. The *io.c* file contains low-level (largely machine-dependent) functions for actually reading characters entered by the user and displaying characters on the screen, and *general.c* contains some low-level utility functions of general use. Finally, the file *main.c* contains the *main* function. This simply calls a number of initialization functions and then calls *do_command*, which first processes the *.str* file and then processes commands entered from the standard input by the user.

APPENDIX **G**

Instructions for Recompiling the PDP Programs

Complete source code for all the programs described in this book is provided in the archive *src.arc*. If you have Version 3.0 (or higher) of the Microsoft C compiler, you should be able to modify the programs and recompile them on your PC. If you have a UNIX system (Berkeley 4.2 or higher) with the standard UNIX C compiler, you should also be able to recompile the programs to run on that system. We first provide a general inventory of the components of the PDP software and of the dependencies among these components. Then we describe the procedure for recompiling for the PC with Microsoft C. Following this we briefly describe how to set up the PDP software on UNIX systems. You are free to try to use other C compilers, but with others you are completely on your own. There is every reason to expect that some tinkering will be required to recompile the software with non-UNIX C compilers other than Microsoft C.

Components of the PDP Software

The software includes seven executable programs: **aa, bp, cl, cs, ia, iac,** and **pa**. For each of these programs, there is a source file with the same name (e.g., *aa.c*). In addition, there are certain other source files that several programs share. The object files that all programs share are grouped into one library file called *libpc.a*. This library file is made up of the compiled versions of the routines from the files *command.c, display.c, general.c, io.c, main.c, patterns.c, template.c,* and *variable.c*. The software also includes the two utility programs: **plot** and **colex**; each of these is constructed from a single corresponding *.c* file.

If you change any source file, you should recompile and relink all object and executable files that depend upon that source file. The following dependency list shows which executables depend upon which source files. Where *libpc.a* is shown, all eight source files in the library are included in the dependency.

aa:	*aa.c, libpc.a*
bp:	*bp.c, weights.c, libpc.a*
cl:	*cl.c, libpc.a*
cs:	*cs.c, weights.c, libpc.a*
ia:	*ia.c, iaaux.c, iatop.c, libpc.a*
iac:	*iac.c, weights.c*
pa:	*pa.c, weights.c, libpc.a*
plot:	*plot.c*
colex:	*colex.c*

Note also that there are many header files (with names ending in *.h*) in the PDP software package. These files often contain declarations that are used in several different modules. This is particularly true for the header files associated with the modules in *libpc.a*. If one of these files is modified, it is prudent to recompile all modules that include this file. The following list indicates which *.h* files are included in each *.c* file:

general.h:	*display.h*
aa.c:	*general.h, aa.h, variable.h, patterns.h, command.h*
bp.c:	*general.h, bp.h, variable.h, weights.h, patterns.h, command.h*
cl.c:	*general.h, variable.h, patterns.h, command.h, cl.h*
command.c:	*general.h, io.h, command.h*
cs.c:	*general.h, cs.h, variable.h, command.h, patterns.h, weights.h*
display.c:	*general.h, io.h, variable.h, template.h, weights.h, command.h*
general.c:	*general.h, command.h, variable.h*
ia.c:	*ia.h, io.h, general.h*
iaaux.c:	*ia.h*
iac.c:	*general.h, iac.h, variable.h, command.h, weights.h, patterns.h*
iatop.c:	*general.h, cs.h, variable.h, command.h, ia.h*
io.c:	*io.h*
main.c:	*general.h, variable.h, command.h, patterns.h*
pa.c:	*general.h, pa.h, variable.h, weights.h, patterns.h, command.h*
patterns.c:	*general.h, command.h, variable.h, patterns.h*
template.c:	*general.h, command.h, variable.h, display.h, template.h*

variable.c:	*general.h, variable.h, command.h, patterns.h, weights.h*
weights.c:	*general.h, command.h, weights.h, variable.h*

Note that all files that include *general.h* also implicitly include *display.h.*

To Recompile for a PC Using Microsoft C

Three batch files are provided in the *src* directory to aid in compiling and linking: *compile.bat, makelib.bat,* and *pdplink.bat.* These files can be executed as though they were programs.

The *compile* batch file includes the command to compile a source file (*.c*) into an object file (*.obj*). The first step in creating an executable (*.exe*) file is to recompile all of the source files upon which it depends, including the files in *libpc.a.* To recompile all the source files, execute the following command in the directory that contains the files:

compile all

This will produce a *.obj* file for each *.c* file in that directory. To compile only one source file, use its name instead of *all* when giving the *compile* command. Thus, to compile *aa.c* you would enter

compile aa.c

This will produce a file called *aa.obj* in that directory. If there are errors in your source code, the compiler may abort the command file and display the error messages.

Once all of the necessary object files are created, the *libpc.a* file can be built. The command file for building the library is executed as follows:

makelib

This will create a file called *libpc.a* in the current directory from the eight object files, which should be in the same directory. Whenever you recompile any of the eight programs in *libpc.a*, you should use the *makelib* command again to update the library. (The *makelib* command requires all the object files to be present, so it is best to keep these files around while you are actively involved in modifying the programs. Once you stop making changes, you can delete the *.obj* files to save space.)

The final stage in compiling is linking. The command file *pdplink.bat* will link the necessary files to create each of the executables. To link the object files for a particular program, enter *pdplink* with the program name as argument. Thus,

pdplink bp

will link the files *bp.obj* and *weights.obj* with the library *libpc.a* to create the executable *bp.exe*. The *pdplink* command will also work with **plot** or **colex**. To create executables for all seven PDP simulation programs, use the following command:

pdplink all

Note that *pdplink all* does not link **colex** and **plot**; these must be linked individually.

If you wish to recompile and relink all of the programs at once, use the following three commands:

compile all
makelib
pdplink all

Once you have created new executables, you will want to move these files into the appropriate working directories, to be used with the relevant *.tem* and *.str* files. The MS-DOS copy utility can be used to do this.

Instructions for Setting Up the PDP Software on UNIX Systems

For UNIX systems, we suggest that you set up a parent directory system for the PDP software and copy the extracted contents of each of the *.arc* files into a separate subdirectory of the parent, giving the subdirectory the same name as the archive. For example, if your parent directory were called */usr/yourname/pdp*, you would put the contents of *aa.arc* into a subdirectory of this directory called */usr/yourname/pdp/aa*. The only files that you will not want to include in this directory system are the *.exe* files, since these will only run on PCs. You would also create a subdirectory called */usr/yourname/pdp/src* containing the source files and other materials necessary to recompile the package from the *src.arc* file.

Once the directories have been set up, you will want to change directories to the *src* subdirectory. It is an easy matter to recompile all the programs because we have supplied a *makefile*. This file is used by the UNIX **make** program to manage the PDP software. To compile all of the PDP simulation programs, you need only execute the following command:

make

To compile a single program, simply give *make* the name of that program as an argument. For example, to recompile the **aa** program, enter

make aa

This form also works with the **plot** and **colex** programs; they are not updated if **make** is executed with no arguments.

In either case, **make** will check the *makefile* to see which source files need to be recompiled and will recompile them. It will update *libpc.a* if necessary. And it will link the necessary object modules together to create the necessary executable files. The supplied makefile places the seven PDP executables in directories that are on the same level as the source directory and have the same name as the executable. For example, if the *src* directory is */usr/yourname/pdp/src* then the **aa** executable would be placed in the directory */usr/yourname/pdp/aa*. If you have set up subdirectories for each program as suggested above, this will all work fine. If you have chosen to organize the directories differently, the *makefile* can be modified to change where each program is placed. For each program there is a variable that specifies the destination directory for the executable version of the program. The names of these variables are uppercase and consist of the program name followed by *DEST*. Thus for **aa**, there is a line in the *makefile* that looks like this:

```
AADEST = ../aa/
```

The path name to the right of the equal sign can be replaced by any other valid UNIX path name. Once it is, **aa** will be stored in the directory specified by the path. Thus

```
AADEST = /usr/foo/pc/bin/
```

would cause **make** to put the **aa** executable in the directory */usr/foo/pc/bin*.

References

Adams, M. J. (1979). Models of word recognition. *Cognitive Psychology, 11,* 133-176.

Anderson, J. A. (1977). Neural models with cognitive implications. In D. LaBerge & S. J. Samuels (Eds.), *Basic processes in reading perception and comprehension* (pp. 27-90). Hillsdale, NJ: Erlbaum.

Anderson, J. A. (1983). Cognitive and psychological computation with neural models. *IEEE Transactions on Systems, Man, and Cybernetics*, *13*, 799-815.

Anderson, J. A., Silverstein, J. W., Ritz, S. A., & Jones, R. S. (1977). Distinctive features, categorical perception, and probability learning: Some applications of a neural model. *Psychological Review*, *84*, 413-451.

Bagley, W. C. (1900). The apperception of the spoken sentence: A study in the psychology of language. *American Journal of Psychology*, *12*, 80-130.

Baron, J., & Thurston, I. (1973). An analysis of the word-superiority effect. *Cognitive Psychology, 4,* 207-228.

Blake, A. (1983). The least disturbance principle and weak constraints. *Pattern Recognition Letters*, *1*, 393-399.

Broadbent, D. E., & Gregory, M. (1968). Visual perception of words differing in letter digram frequency. *Journal of Verbal Learning and Verbal Behavior*, *7*, 569-571.

Carr, T. H., Davidson, B. J., & Hawkins, H. L. (1978). Perceptual flexibility in word recognition: Strategies affect orthographic computation but not lexical access. *Journal of Experimental Psychology: Human Perception and Performance,* 4, 674-690.

Cattell, J. M. (1886). The time taken up by cerebral operations. *Mind, 11,* 220-242.

Feldman, J. A. (1981). A connectionist model of visual memory. In G. E. Hinton & J. A. Anderson (Eds.), *Parallel models of associative memory* (pp. 49-81). Hillsdale, NJ: Erlbaum.

Fukushima, K. (1975). Cognitron: A self-organizing multilayered neural network. *Biological Cybernetics*, *20*, 121-136.

Geman, S., & Geman, D. (1984). Stochastic relaxation, Gibbs distributions, and the Bayesian restoration of images. *IEEE Transactions on Pattern Analysis and Machine Intelligence*, *6*, 721-741.

Grossberg, S. (1976). Adaptive pattern classification and universal recoding: Part I. Parallel development and coding of neural feature detectors. *Biological Cybernetics*, *23*, 121-134.

Grossberg, S. (1978). A theory of visual coding, memory, and development. In E. L. J. Leeuwenberg & H. F. J. M. Buffart (Eds.), *Formal theories of visual perception*. New York: Wiley.

Grossberg, S. (1980). How does the brain build a cognitive code? *Psychological Review*, *87*, 1-51.

Hebb, D. O. (1949). *The organization of behavior*. New York: Wiley.

Hinton, G. E. (1977). *Relaxation and its role in vision*. Unpublished doctoral dissertation, University of Edinburgh.

Hinton, G. E., & Anderson, J. A. (Eds.). (1981). *Parallel models of associative memory*. Hillsdale, NJ: Erlbaum.

Hinton, G. E., & Sejnowski, T. J. (1983). Optimal perceptual inference. *Proceedings of the IEEE Computer Society Conference on Computer Vision and Pattern Recognition*, 448-453.

Hopfield, J. J. (1982). Neural networks and physical systems with emergent collective computational abilities. *Proceedings of the National Academy of Sciences, USA*, *79*, 2554-2558.

Hopfield, J. J. (1984). Neurons with graded response have collective computational properties like those of two-state neurons. *Proceedings of the National Academy of Sciences, USA*, *81*, 3088-3092.

James, W. (1890). *Principles of psychology* (Vol. 1). New York: Holt.

Johnston, J. C. (1978). A test of the sophisticated guessing theory of word perception. *Cognitive Psychology, 10,* 123-153.

Johnston, J. C. (1980). Experimental tests of a hierarchical model of word identification. *Journal of Verbal Learning and Verbal Behavior, 19,* 503-524.

Johnston, J. C., & McClelland, J. L. (1973). Visual factors in word perception. *Perception & Psychophysics, 14,* 365-370.

Johnston, J. C., & McClelland, J. L. (1974). Perception of letters in words: Seek not and ye shall find. *Science, 184,* 1192-1194.

Johnston, J. C., & McClelland, J. L. (1980). Experimental tests of a hierarchical model of word identification. *Journal of Verbal Learning and Verbal Behavior, 19,* 503-524.

Jordan, M. I. (1986). Attractor dynamics and parallelism in a connectionist sequential machine. *Proceedings of the Eighth Annual Meeting of the Cognitive Science Society*. Hillsdale, NJ: Erlbaum.

Kernighan, B. W., & Ritchie, D. M. (1978). *The C Programming Language*. Englewood Cliffs, NJ: Prentice-Hall.

Kohonen, T. (1977). *Associative memory: A system theoretical approach*. New York: Springer.

Kucera, H., & Francis, W. (1967). *Computational analysis of present-day American English*. Providence, RI: Brown University Press.

Levin, J. A. (1976). *Proteus: An activation framework for cognitive process models* (Tech. Rep. No. ISI/WP-2). Marina del Rey, CA: University of Southern California, Information Sciences Institute.

Luce, R. D. (1959). *Individual choice behavior*. New York: Wiley.

Manelis, L. (1974). The effect of meaningfulness in tachistoscopic word perception. *Perception & Psychophysics, 16,* 182-192.
Massaro, D. W. (1973). Perception of letters, words, and nonwords. *Journal of Experimental Psychology, 13,* 45-48.
Massaro, D. W., & Klitzke, D. (1979). The role of lateral masking and orthographic structure in letter and word recognition. *Acta Psychologica, 43,* 413-426.
McClelland, J. L. (1976). Preliminary letter identification in the perception of words and nonwords. *Journal of Experimental Psychology: Human Perception and Performance,* 2, 80-91.
McClelland, J. L. (1979). On the time-relations of mental processes: An examination of systems of processes in cascade. *Psychological Review, 86,* 287-330.
McClelland, J. L. (1981). Retrieving general and specific information from stored knowledge of specifics. *Proceedings of the Third Annual Meeting of the Cognitive Science Society,* 170-172.
McClelland, J. L., & Johnston, J. C. (1977). The role of familiar units in perception of words and nonwords. *Perception & Psychophysics, 22,* 249-261.
McClelland, J. L., & Rumelhart, D. E. (1981). An interactive activation model of context effects in letter perception: Part 1. An account of basic findings. *Psychological Review, 88,* 375-407.
McClelland, J. L., & Rumelhart, D. E. (1985). Distributed memory and the representation of general and specific information. *Journal of Experimental Psychology: General, 114,* 159-188.
McClelland, J. L., Rumelhart, D. E., & the PDP Research Group. (1986). *Parallel distributed processing: Explorations in the microstructure of cognition. Vol. 2. Psychological and biological models.* Cambridge, MA: MIT Press/Bradford Books.
Minsky, M., & Papert, S. (1969). *Perceptrons.* Cambridge, MA: MIT Press.
Morton, J. (1969). Interaction of information in word recognition. *Psychological Review, 76,* 165-178.
Norman, D. A., & Bobrow, D. G. (1975). On data-limited and resource-limited processes. *Cognitive Psychology, 7,* 44-64.
Pillsbury, W. B. (1897). A study in apperception. *American Journal of Psychology, 8,* 315-393.
Pinker, S., & Prince, A. (1987). *On language and connectionism: Analysis of a parallel distributed processing model of language acquisition* (Occasional Paper 33). Cambridge: Massachusetts Institute of Technology, Center for Cognitive Science.
Reicher, G. M. (1969). Perceptual recognition as a function of meaningfulness of stimulus material. *Journal of Experimental Psychology, 81,* 274-280.
Riley, M. S., & Smolensky, P. (1984). A parallel model of (sequential) problem solving. *Proceedings of the Sixth Annual Conference of the Cognitive Science Society.*
Rosenblatt, F. (1959). Two theorems of statistical separability in the perceptron. In *Mechanisation of thought processes: Proceedings of a symposium held at the National Physical Laboratory, November 1958. Vol. 1* (pp. 421-456). London: HM Stationery Office.
Rosenblatt, F. (1962). *Principles of neurodynamics.* New York: Spartan.
Rumelhart, D. E. (1977). Toward an interactive model of reading. In S. Dornic (Ed.), *Attention & Performance VI.* Hillsdale, NJ: Erlbaum.
Rumelhart, D. E., & McClelland, J. L. (1982). An interactive activation model of context effects in letter perception: Part 2. The contextual enhancement effect

and some tests and extensions of the model. *Psychological Review*, *89*, 60-94.

Rumelhart, D. E., McClelland, J. L., & the PDP Research Group. (1986). *Parallel distributed processing: Explorations in the microstructure of cognition. Vol. 1. Foundations.* Cambridge, MA: MIT Press/Bradford Books.

Rumelhart, D. E., & Ortony, A. (1977). The representation of knowledge in memory. In R. C. Anderson, R. J. Spiro, & W. E. Montague (Eds.), *Schooling and the acquisition of knowledge* (pp. 99-135). Hillsdale, NJ: Erlbaum.

Rumelhart, D. E., & Siple, P. (1974). Process of recognizing tachistoscopically presented words. *Psychological Review, 81,* 99-118.

Rumelhart, D. E., & Zipser, D. (1985). Feature discovery by competitive learning. *Cognitive Science*, *9*, 75-112.

Selfridge, O. G. (1955). Pattern recognition in modern computers. *Proceedings of the Western Joint Computer Conference.*

Smolensky, P. (1983). Schema selection and stochastic inference in modular environments. *Proceedings of the National Conference on Artificial Intelligence AAAI-83*, 109-113.

Spoehr, K., & Smith, E. (1975). The role of orthographic and phonotactic rules in perceiving letter patterns. *Journal of Experimental Psychology: Human Perception and Performance*, *1*, 21-34.

Turvey, M. (1973). On peripheral and central processes in vision: Inferences from an information processing analysis of masking with patterned stimuli. *Psychological Review*, *80*, 1-52.

von der Malsberg, C. (1973). Self-organizing of orientation sensitive cells in the striate cortex. *Kybernetik*, *14*, 85-100.

Weisstein, N., Ozog, G., & Szoc, R. (1975). A comparison and elaboration of two models of metacontrast. *Psychological Review*, *82*, 325-343.

Wheeler, D. D. (1970). Processes in word recognition. *Cognitive Psychology, 1,* 59-85.

Whittlesea, B. W. A. (1983). *Representation and generalization of concepts: The abstractive and episodic perspectives evaluated.* Unpublished doctoral dissertation, MacMaster University.

Widrow, G., & Hoff, M. E. (1960). Adaptive switching circuits. *Institute of Radio Engineers, Western Electronic Show and Convention, Convention Record*, *Part 4*, 96-104.

Index

Notes for Using The PDP Software on Macintosh Computers

J. L. McClelland and D. E. Rumelhart

The two diskettes enclosed with this document contain a version of the PDP software for use on Macintosh computers. The original version of the software is described in *Explorations in Parallel Distributed Processing: A Handbook of Models, Programs, and Exercises* by J. L. McClelland and D. E. Rumelhart (Cambridge, MA: MIT Press, 1988). The Macintosh version of the software functions virtually the same as the original version. It uses a 24x80 character window, and input and output are entirely character-based. The slight differences between the two versions are detailed herein. License to use the Macintosh version of the software is granted to anyone who wishes to use it under the terms of the license agreement, which is reprinted at the end of this document. The license agreement may also be found in the *src* folder on PDP Disk 2.

Purchasers should be aware that this version of the software is *not* suitable for porting to non-Macintosh systems. The original version of the software is more suitable for that purpose. Macintosh owners who need to purchase this Mac version for use on the Macintosh but would like to port the software to another computer system as well should contact PDP Software Inquiries, at the address given at the end of the *Trouble Shooting* section.

WHAT IS INCLUDED

The software package comes on two diskettes. Disk 1 contains a folder for each program and a folder for the two utility programs, **colex** and **plot**. Each program folder is ready for immediate use. Disk 2 contains the source code and the materials necessary for recompiling the programs under THINK's Lightspeed C (see below), as well as versions of the programs that can run on Macintosh computers that have the 68881 or 68882 coprocessor.

WHAT YOU NEED

The software runs on Macintosh Plus, Macintosh SE, and Mac II computers. A double-sided floppy drive is required to read the diskettes and a hard disk is strongly recommended. You must be running current versions of the system and finder or multifinder. You must also have the **Courier 10** font. (This font is supplied with all Macintosh systems.) The 68881 versions of the programs require either the 68881 or the 68882 math coprocessor. They are many times faster than the standard versions.

You will need a copy of the PDP Handbook, which contains instructions on how to use the programs. You will also need a text editor. A simple one should be sufficient. A graphics application like **CricketGraph** would be very useful for making graphs and plots, though very rudimentary plotting capability is provided via the **colex** and **plot** programs.

The information that follows assumes that you have a working knowledge of the Macintosh. (This can be obtained from the *Apple System Software User's Guide.*)

Older, Smaller Macs

The software does run (slowly) on some 512e's, but has not been thoroughly tested on these machines. If the system and finder are reasonably recent (5.3 seems to work), and you have a double-sided drive (along with a hard disk or second drive for the system), you should be ok. You must also make sure that the **Courier 10** font is installed on your system. If you're stuck with single-sided drives, ask a friend with a double-sided drive to split up the software onto single-sided diskettes for you.

USING THE PROGRAMS

You can use the software right off of *PDP Disk 1* if you wish, but there is no extra space on this disk for saving results. If you have a hard disk, it would make sense to make a new folder (let's say you call it PDP) and copy all the folders from *PDP Disk 1* into it. If you will be working from floppies, it makes sense to put just one or two of the program folders onto each diskette.

If you do have a math coprocessor, you will probably want to replace the executable file in each program folder with the 68881 version located in the *68881* folder on *PDP Disk 2*. You can either delete the non-68881 executables from the program folders or just rename them. Then copy the individual executables from the *68881* folder on Disk 2 into each of the corresponding program folders.

If you wish, you can also put the *src* folder located on PDP Disk 2 into your PDP folder. Another place to put it is into a folder that encompasses your Lightspeed C compiler.

You're almost ready to start using the PDP software. Open up the program folder of your choice, then just double-click on the application icon to start the program. The program will open a window and prompt you for a template file and a startup file (these are described in chapter 2 of the PDP Handbook, page 24). Enter the names of the files you want the program to use. Thus, for the first exercise in chapter 2, you would want to enter *jets.tem* as the template file and *jets.str* as the startup file.

From this point on, the use of the programs on a Mac is virtually identical to their use on the PC as described in the PDP Handbook. There are these minor differences:

1. File names must refer to files in the same directory as the application program, or must specify full path names beginning with the drive name. (You may of course make many copies of each application.) Unfortunately spaces are interpreted as word boundaries by the PDP programs, so path names that include spaces will not work.

2. The *run* command, which allowed execution of other processes from inside the PDP programs, is not available on the Mac. The programs are MultiFinder-compatible, however, and can be run in the background while you do other things.

3. To interrupt the programs, you must use Command-c or Command-., rather than Control-c. (Command is the key to the left of the spacebar with the little square with loops at the corners. To type Command-c you hold down Command while typing c.)

4. All floating-point arithmetic is carried out in double precision on the Mac. Using double precision, we have found that the programs produce results that are identical to those produced on other computers.

5. The Mac version of the **ia** program must read in a list of words and resting activation levels; in other versions this list is compiled directly into the program. The list is in the file *words.lis* and is read in through a command in the *ia.str* file.

6. In the Mac version the entries in log files are separated by tabs, rather than

spaces as in the PC version. This makes the log files directly interpretable to programs such as **CricketGraph** and other Mac programs that expect fields to be tab-separated.

Menu and Finder Issues

When a PDP program is foregrounded, there will be three menus available in the menu bar: The **Apple** menu, the **File** menu, and the **Edit** menu. The **Apple** menu will contain the set of tools that are typically available on your system under this menu. The **File** menu contains two options, *Close* and *Quit*. The **Edit** menu contains a few standard edit options such as *Cut*, *Paste*, etc. With the exception of *Quit*, the menu options under the **File** and **Edit** menus are inoperative while the PDP program is in focus. Some of them are activated, however, under some of the tools that are available under the **Apple** menu. For example, when using the *Notepad* tool in the **Apple** menu you can cut and paste.

Clicking on *Quit* in the **File** menu (or entering Command-Q) causes the program to quit immediately. You can also quit the normal way, by entering *quit* to the program and then confirming with *y*.

If you are running the PDP software under the MultiFinder, you will be able to switch the focus between applications in the normal way. The PDP software will keep running in the background under the MultiFinder, so you can, for example, start a simulation going, then open up a text editor and work on a document while the program is running.

PDP program windows can be resized using the sizing icon in the lower left corner, and they can be moved using the title bar, but they do not scroll.

TROUBLE SHOOTING

If you encounter difficulty reading the diskettes, make sure you are using a two-sided drive. If you still have problems, the diskette is probably defective and should be returned for replacement (see the end of this section for address information).

If a program crashes with a Macintosh system error message before you are prompted for any input, you may be running the 68881 version on a machine without a coprocessor. Alternatively, you may be running the standard version on an early Mac 512 with which it is not compatible, or your system and/or finder may be out of date; this problem has occurred on Mac SE's under system and finder versions 4.2, for example.

If you are running a large network or are using a large number of patterns, the program may exceed the memory that is allocated to it, and give the error message "Out of memory" during start up. If this happens, get back to the Finder or MultiFinder, select (click once on) the executable program file, then click on the *Get Info* option in the **File** menu, and increase the value in the *Application Memory Size* field in the lower right hand corner of the Dialog Box that pops up. Then try to run your program again.

If you increase the Application Memory Size (or even if you don't), there is a chance that a program will ask for more memory than is available on your system when it tries to run. The Macintosh system will report this, and may or may not let you try to run with less than the full amount specified in the Application Memory Size. If the system will not let you run, or if the program then crashes, killing other running processes and/or reducing the Application Memory Size can help. If not, the only solution is to buy more memory.

If the software does not crash but you see a jumbled mess of character fragments instead of a legible screen display, it probably means that you need to install the **Courier 10** font. This font is distributed with all Macintosh systems, but it may have been deleted from your system. Information on how to reinstall this font can be found in the *Macintosh Utilities User's Guide*.

If the software runs but you encounter problems running it with *.str*, *.net*, and/or *.tem* files that you have created or modified, there are many errors that you may have made. Often these are just simple syntactic errors that can be corrected by working through the *.net*, *.str*, and *.tem* files that your are using. The following two common problems are a bit subtler:

You may not have provided the information needed to initialize the network early enough in the *.str* file. For **iac**, **cs**, **pa**, and **bp**, the *get/ net* command should be at the beginning of the *.str* file. For **aa**, the *get/ net* command is not used, but you need to set the number of units with the *set/ conf/ nunits* command at the beginning of the *.str* file. For **cl**, you need to set the number of inputs and the number of outputs (using *set/ conf/ ninputs* and *set/ conf/ noutputs*). These commands must be executed before the program can read in patterns, update the screen, or perform any actual network computations.

It should also be noted that when specifying blocks of weights using the *%r* notation in *.net* files, *each receiving unit can only be specified in one such specification.* See Appendix C of the Handbook, pages 267–268, for further discussion.

If after reading this you still have a problem, please write to PDP Software Inquiries, c/o Texts Manager, MIT Press, 55 Hayward Street, Cambridge, MA 02142. If you have a defective diskette, send it along and it will be replaced. If the software will not run satisfactorily on your system, you may return your purchase for a refund. For other problems, we cannot offer help on an individual basis, but if you enclose a stamped, self-addressed envelope, you will receive an acknowledgment along with accumulated advice, fixes, work-arounds, and information on availability of subsequent releases. Comments and suggestions for improvement will be appreciated as well.

UTILITIES

Two very simple utility programs, **colex** and **plot**, are provided and function nearly as described in the Handbook. These utilities are very rudimentary, and there are plenty of Macintosh programs such as **CricketGraph** that are much better. We strongly recommend that you get such a program if you want to draw graphs reflecting the changing value of one or more variables over time. **Colex** may remain useful for pulling out selected columns from larger log files, though you might try to get in the habit of learning to control logging so that only the stuff you really want to plot gets logged (see the PDP Handbook, Appendix C and the beginning of Appendix D).

Colex and **plot** can only access files in the directory from which they are run. The arguments these programs require are given in a window that pops up after the program is fired up. Thus, for example, to create the plot shown in the PDP Handbook, Appendix D, page 287, you would first create the *jets.log* file (see pages 283–284). Then perform the following steps:

1. Open the *utils* folder in your PDP folder.

2. Create duplicates of **colex** and **plot** in the *iac* subfolder. (Use the duplicate command under the (Multi)Finder's **File** menu. You might lose track of the programs if you move them.)

3. Double-click **colex**. When prompted, enter the arguments required to tell the program what you want to do (page 285):

   jets.log jets.clx 0 6 15 31

   When this is done, click on the exit box to leave **colex**.

4. Double-click on **plot**. When prompted, enter the command line arguments for this program (page 287):

   jets.fmt jets.clx

   The plot will then be displayed on the screen. To save the plot in a file, just add a filename argument.

WORKING WITH THE SOURCES CODE

The Macintosh versions of the PDP programs were compiled under THINK's Lightspeed C (Symantec Corporation, THINK Technologies Division, 135 South Road, Bedford, MA 01730; phone (617) 275-4800). Some work would be required to recompile them under another C compiler. **We have no experience with other compilers and can offer no assistance.** Here we explain briefly what has been added to permit the programs to run on Macintosh computers, and we describe what you would need to do to recompile them under Lightspeed C. We assume familiarity with C and the Lightspeed C compiler.

Additions for the Mac

Two sets of routines had to be added to the PDP software for use on the Macintosh. These are in the files *el.c* and *curses.c*. The name *el.c* stands for event loop, and this file contains routines that are responsible for interfacing the PDP software with the Macintosh architecture. The file *curses.c* contains routines that create the 24x80 character screen emulation window and read and write characters to and from the screen, emulating parts of the BSD UNIX *curses* library.

In addition to these two sets of routines, the Mac requires a *.rsrc* file for each PDP program. The *.rsrc* file specifies the contents of the menus and a few other resources that are available to the programs.

The file *mac_stuff.h* contains declarations that are specific to the Macintosh (and to Lightspeed C). This is included in *general.h*, which in turn is the first file included in every *.c* file in the entire package.

There have been various internal changes and slight rearrangements of the code for use on the Macintosh. The most important change is the use of double-precision floating-point arithmetic throughout.

Recompiling the Programs

Under Lightspeed C, each program is associated with a project. The source folder contains a project file for the **bp** program. For the other programs, you can just duplicate the bp project, rename it, and adapt it as follows. For **cs**, **iac**, and **pa**, you need simply replace *bp.c* with *cs.c*, *iac.c* or *pa.c*. For **aa** and **cl**, you would replace *bp.c* with *aa.c* or *cl.c*, and also remove *weights.c*. For **ia**, you would replace *bp.c* and *weights.c* with *ia.c, iaaux.c* and *iatop.c*.

The source folder also contains a resource file for **bp**, *bp.rsrc*. To create the resource files for the other programs, make a copy of *bp.rsrc*, name it to go with the other program (e.g., *cs.rsrc*), and then, using **ResEdit**, click down into DLOG=256 of the new *.rsrc* file and then into the Dialog Box template itself and change the entry in the upper left corner to the new program's name. If this turns out not to work, you'll just be stuck with the name of the wrong program in the Dialog Box that pops up when you click on the "About xx" item (where xx is the name of the program) in the Apple menu.

In addition, the source folder contains a subfolder called **pdplib** that contains the **pdplib** project. This project creates the *lpdp* library, which is included in each of the PDP programs. This folder contains the *max_stuff.h* include file and the other general-purpose include files as well as the library sources.

To recompile, you must first rebuild the library, then the applications. Rebuilding is very simple. First open the **lpdp** project, and click on *Build Library* in THINK C's **Project** menu. You'll be asked to specify a filename for the new version of the library. It makes sense to use the old name, **lpdp**, since this is what all the programs expect. After you've rebuilt the library, open the project for the program you want to rebuild, and then just click on *Build Application*. You will be asked to specify a destination filename.

To recompile the programs for use with the 68881, you must follow these steps before recompiling:

1. Edit *mac_stuff.h* in the *pdplib* directory, and remove the comments around

   #define _MC68881_

   Then, add comments around

   #define _ERRORCHECK_

2. For each project:

   a. Turn on the 68881 code generation option using the *Options* item on the **Edit** menu.

   b. Replace the *Math*, *stdio*, and *unix* libraries in each project with their counterparts: *Math881*, *stdio881*, and *unix881*. These are located in the Libraries folder of the THINK C Folder. These three libraries, together with the *storage* library, should occupy the second segment of the project.

Note that (a) applies to the **pdplib** project as well as to each application project; (b) applies only to the application projects.

Bugs in Lightspeed C

The Lightspeed C compiler (Version 3.0) that was used to create these programs is not without problems. (a) In the distributed *MathHybrid* Library neither *_pow* nor *_exp* returned a result to the calling routine. We use the

Math881 library with satisfactory results. (b) The documentation fails to specify that _ERRORCHECK_ must not be defined when using the Math881 library. (c) An apparent compiler error had to be circumvented by rearranging the order in which subroutines occurred within a source file. (d) Finally, the **ResEdit** (Version 1.1b3, an Apple product) supplied with THINK C occasionally trashes the screen, although it seems to work if you just click where things should be even if there is trash there. Version 1.2b2 appears to work better.

ACKNOWLEDGMENTS

Ross Thompson did the initial port of the PDP software to the Macintosh and wrote the first version of *el.c* and *curses.c*. Gregory Fox made the programs MultiFinder-compatible, and added resizing and window movement support. Greg's efforts were supported by the College of Humanities and Social Sciences at Carnegie-Mellon University. We thank both Ross and Greg for their contributions to making the PDP software more widely accessible.

License Agreement for the Parallel Distributed Processing Software Package

The software distributed on the two diskettes *PDP Disk 1* and *PDP Disk 2* constitute the Macintosh version of the PDP software package.

CONDITIONS OF LICENSE. This software is licensed to users under the following terms:

1. The licensee may use this software and modified or extended versions thereof, and may make copies of the source and object code to support this use, provided (a) that any reports or publications arising from such uses acknowledge the source of the software and refer to the Handbook; (b) that the licensee acknowledges and accepts the disclaimers given below; and (c) that the licensee agrees to refrain from marketing or distributing for profit any portion of the software or any modification or extension thereof.

2. The licensee may make modifications and extensions of the software, as long as (a) all versions of the source files used for such purposes still retain the original copyright notice; (b) the computer-readable version of this license agreement in the file *license.txt* is kept in the same directory with these source files; and (c) each executable program created from such modification or extension displays the original copyright notice each time it is run.

This license is granted to the purchasing individual or institution, and to all individuals who are given access to the software by the purchaser or purchasing institution. The software may also be used and adapted in accordance with these terms for any purpose of the United States government.

WARRANTY ON DISKETTES. The MIT Press warrants to the original purchaser only that the diskettes on which the PDP software is supplied are free from defects as supplied; The Press will replace any defective diskettes free of charge.

DISCLAIMERS. The programs have been carefully tested, but they are provided without any warranty other than that stated above. There is no warranty that the programs will produce mathematically correct results or that they are free from errors. No warranties are made concerning the conditions under which the source code can be recompiled, nor about the specific system configurations under which the software will or will not run. However, if the purchaser is unable to make any use of the software after the information in the *Trouble Shooting* section of the *Notes for using the PDP software on Macintosh computers* has been taken into account, the purchaser may return the product purchased to the address given in the *Trouble Shooting* section for a refund.